Quantum Void: the Mystery of Empty Space

Dedication

This book is dedicated to my wife Sara, my Irish colleague and life partner, who has enriched my life abundantly.

CONTENTS

Foreword

A funny thing happened one morning at a breakfast bar in Boise, Idaho. I had recently retired from employment at a large high-tech corporation, where we were involved in state-of-the-art R&D work in signal processing. My mind was still ringing from the speed at which my life was changing. The situation left me feeling both free and vacant. After all, I had been doing this for more than forty years.

Nevertheless, something was pushing me to write: write about what? The pen is full of ink, the paper is clean and blank, so what is my rub? Staring into blank space, I felt the nudge; it became clear: write about your favorite activity. Write about physics!

So, I began to write without a plan and a clear objective. I wrote and drank hot coffee until I was expended. I wrote about physics. I wrote about mathematics. One might think I would write about new theoretical techniques or experimental techniques, but the writing was not about the mechanics and new theories or recent experimental discoveries. Instead, it was about the physics environment- what it is like being a physicist.

This book was written to shed light on the recent observations that new physics students with academic promise have been unexpectedly walking away from the courses for no known reason. The American Physical Society recently announced that this is happening. Based on my observations of events like this, I began my version of a book about physics that leads the student through a pathway of discussions and observations about physics. The text itself is not focused on the physics; instead, it tells, and in some cases intertwined, a few stories for the reader. This storytelling has the effect of educating the student on what it is like to be a physicist. It appears this approach has the dual effect of teaching some physics to the student without heavy mathematics while fortifying his/her confidence to take on higher-level physics.

So, hopefully, if it has resulted in you being excited about physics, then it has achieved its purpose.

Preface

When I was an eight-year-old boy, my family purchased our first television set. It was big and clunky; a beautiful thing indeed! We could watch shows together in the evening without leaving the comfort of home. I was exposed to new places and things as never before.

But there was one special thing I was exposed to through that TV set. Every Saturday morning at 6:00, while the family slept, I would rise, quietly go downstairs to the living room and turn on the TV. I would sit in silence and watch the TV test pattern as I waited. Then, suddenly the picture would change and the show I waited for would begin. It was called "Mr. Wizard", a half-hour discussion of science. Mr. Wizard (Don Herbert) would appear on the screen and say something like "Good morning boys and girls; today we are going to learn about rockets and how they work. Here are the things you will need", and he would list out some common items we could get from the kitchen. Then we would do the experiment. I would follow along as Mr. Wizard showed me how things work.

I didn't realize it at the time, but as I sat in front of the TV waiting, I watched, followed along and experienced these little experiments, I was being exposed to the world of scientific discovery. I was quietly learning some very important lessons about science. First of all, I was learning that science is not about answers, but questions. The idea that science is all about facts and answers is very unfortunate. Our world demands quick answers to complex questions. Scientists who are exposed to these complex questions learn that the answers don't always come quickly. Often, we find that for answers to emerge, the question must be stated differently.

As a physicist, pondering and understanding some of nature's complexities has been my life's work. My interest and curiosity in how things work and why they happen began when I was quite young. Late one evening, when I was seven years old, my parents awakened me to view the aurora borealis or northern lights. I remember observing the beauty of the many colors in the night sky and wondering what it was in nature that made this showing possible.

In writing this book, I want to:

- Spark the interest of budding scientists to get serious and pursue science. If I am successful, I will have conveyed the sense of wonder and awe that I and many scientists experience when nature reveals herself through science.

- Demonstrate that someone uninitiated in mathematics can get shanghaied into the world of science.

- Describe science as a great adventure, a never-ending story that compels at the highest levels.

- Create for the reader what a strong motive force is that drives scientists to understand our universe better.

I have introduced the terms Big Space and Small Space alternatively in the text. The technology accompanying the two is quite different, but it is blending over time.

I hope you enjoy reading this book and that you find the reading triggers the most important emotional reaction: the desire to learn.

Interestingly, the number of undergraduate students majoring in physics has increased in the last decade. However, the attrition rate is very high for those who pack the physics classrooms initially when only a fraction of those students go on to major in a physics degree. "Too many switch majors to similar programs like engineering or computer science. What gives?".

In the October 2024 American Physical Society (APS) News issue, the lead article was "Introductory Physics Classes Can Make or Break Students' Persistence in the Field." One key finding in this five-year longitudinal study, beginning in 2018, found that "A college student's first physics course marks a key stage in a budding career." One major finding was that a critical success factor in majoring in physics was that if the student "comes in with confidence, then they are more likely to do well. In other words, "mindset has a huge impact."

Chapter 1 - Space

Big-Space and Small-Space

Scientific Method

Our universe is filled with empty space. That does not mean our universe is empty. In fact, as far as we can tell, empty space is somehow not so empty. The past hundred years of physics have led the community in directions never imagined by even the brightest among us. Not only that, but there is no reason to believe that the journey is anywhere near complete.

Introduction: Big Space and Small Space

The universe is a really big place; we all know that. When we look into the night sky, we see a blanket of stars everywhere. But there is a lot of black out there, too. If stars were really everywhere, the night sky would be white, not black! In actual fact, the most abundant part of the night sky is empty space.

In our galaxy, the Milky Way, there are approximately one hundred billion stars. That's a lot of stars, but there is even more space. For each star in our galaxy, there are about 25 trillion miles of empty space between stars. Amazingly, way more than 99.9999999% of our universe is empty space; actually, the number is more like 99.999999 plus at least 30 more nines!

So, space fills the void that is our universe. It fills the vast expanse between planets, stars, galaxies, as well as molecules, atoms and sub-atomic particles that make up our world. If instead of looking to outer space, we look deeply into matter at "inner space" in terms of molecules and atoms, we find a similar story. An atom consists of a central nucleus surrounded by a "cloud" of a few electrons. But the atom's diameter is 100,000 times larger than the diameter of its nucleus; that nucleus is but a tiny pinpoint in the center of a "cloud" of our new friend, empty space! So, in fact, just like outer space, inner space is overwhelmingly empty space.

Let's look at Figure 1-1 to get a better idea of the size of our universe.

Figure 1-1: Our universe is both very large and very small. In science, we sometimes refer to the very small aspect of our universe as "reciprocal space" because smallness is inversely related to bigness. These concepts are often exploited mathematically in physics, to help us "transform" from the very large to the very small.

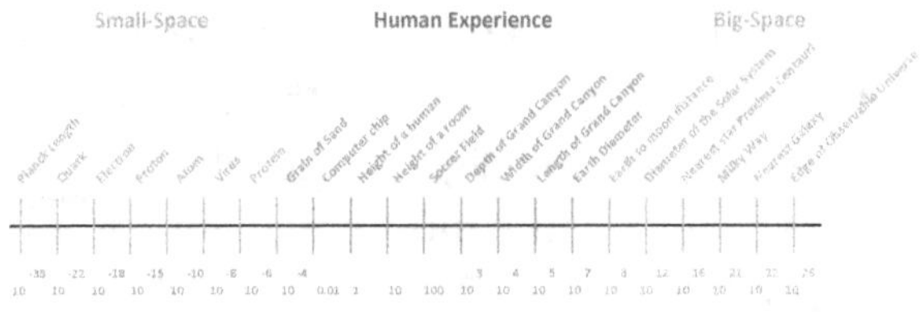

The expanse is very great compared to the size of things we encounter in everyday life. The smallest objects we typically encounter in our life are the size of a grain of sand; maybe one tenth of a millimeter, or 10-4 meters.

Scientists often work with very large or very small numbers. When they do, they use a shorthand technique called *"sicin-tific notation"*. It utilizes the powers of ten which 1,000 s 10 x 10 x 10, so we can write it as 103. If the number is 2,000, we can write it as 2×10^3. Or if the number is nine billion, which is usually written as 9,000,000,000, we can write it more succinctly as 9×10^9. Similarly, we can write a very small number like 0,00000004 as 4×10^{-8}.

Scientific Notation

The largest object in everyday life is our earth, which is about 4,000 miles, or about 13 million meters in diameter.

So, sizes in our everyday life range from 10^{-4} meters to 10^{+7} meters. That's eleven factors of ten, or in scientific language, eleven orders of magnitude. That's a pretty wide range of sizes, but it is very tiny in comparison to the range of sizes making up our universe.

In our universe, there are the very big and the very small. From here on, we will call these two size distinctions "Big-Space" and "Small-Space." Starting with Earth as the largest object we deal with in everyday life and looking at larger objects, we see from the figure that the farthest out into space that we can see, so far, is 1026 meters from where we are. If we could string a bunch of Earth-sized planets together like pearls, it would take ten million trillion Earths to span the distance from here to the edge of our universe! It takes light more than ten billion years to reach us from that distance. These numbers are truly "astronomical"! Getting a realistic feeling for the meaning of these large numbers is nearly impossible for mortals like us.

But in its unique way, "Small Space" is every bit as impressive as "Big Space." That grain of sand we were talking about is made of atoms of silicon. Each of those atoms is a million times smaller, yet at 10-^{10} meters. Each silicon atom has a nucleus that is made of protons and neutrons. Each of those is 10^{-15} meters small, another hundred thousand times smaller than the atom, and it continues to the smallest length that makes any sense with the physics we know today, the "Planck length", at 10^{-35} meters.

So, it looks like empty space spans dimensional sizes from the largest at 10^{26} meters to the smallest known at 10^{-35} meters; that's 61 orders of magnitude at least, spanning our universe. At every turn in this span, we find that the volume of our universe is overwhelmingly occupied by empty space. Is this "much ado about nothing"?

An important question now being asked by the scientific community is, "Is space really empty?" Is space just the emptiness between atoms or stars or galaxies? Or is there something in empty space that so far eludes us? When scientists speak of "curved space," what do they really mean? Does space have a structure we cannot yet detect? If so, how could we detect it? How is that structure consistent with the laws of Physics that we already understand?" And, even more importantly, "Are there tenable theories, based on science we do understand, that could guide us in our search for answers?" Such questions are the heart of physics today, and they are more easily asked than answered.

The scientific community's understanding of empty space has been unfolding like a thriller mystery novel for hundreds of years. Even today, it continues; chapter by chapter, event by event, conflict and resolution, rising action; it's all there! Every time we close a chapter with some answers, a new chapter opens with a whole new set of questions. Nobody knows where it will end, how it will end, or even if it will ever end! Perhaps YOU will write the next chapter!

On Scientific Method

As we take our journey of the science of empty space, we will learn what it is like to be "doing physics" or, more generally, “doing science”. One of the most important and useful tools for doing science is the Scientific Method. The notion of the Scientific Method is based on the idea that nature is not fickle or ambiguous. That is, any physical phenomenon detected and measured at one place and time will be measurable at another place and time if the conditions are the same. Therefore, this concept of measurement reproducibility is held as one of the most important criteria for verifying measurement validity. In plain words, if someone has an idea of how a particular observed physical phenomenon occurs, there is a process that helps the scientist pursue this idea and reach a rational conclusion on whether the idea has merit.

The Scientific Method is an orderly process tool that the scientist can use to navigate our way through understanding of nature. While this process is orderly and rational, the scientist’s ideas need not be. The Scientific method will help the scientist develop their ideas. The idea, or theory, need not be based on logic. It might be an intuitive notion or “gutfeel” that is fed by experience in the lab. Or it might be some wild idea that flies in the face of everything science knows about the subject. But none of that matters; the Scientific Method will work as well on a “bad” idea as on a “good” idea. If an idea is baseless or irrational, application of the Scientific Method will reveal these issues.

Most textbooks on science will present the Scientific Method as a “list” of things that happen as a scientist pursues an idea of how nature works. It might look like this.

- Observe a phenomenon, in the lab or in the field, to establish what nature does.
- From these observations, formulate a theory to explain what is observed.
- Design and implement an experiment or set of experiments that will probe and test the validity of the theory.
- Analyze the data accumulated in the experiment, toward comparing measured results to theorized results.
- Compare experimental results with theoretically predicted results.
- Reach a conclusion on the validity of the theory in explaining the observed results.

There are a few guidelines and criteria for doing high quality scientific research:

A scientific result must be reproducible. This means that multiple independent measurements of the same physical quantity under the same conditions should yield the same result.

It also means that **measurements of the same physical quantity using different measurement techniques (by different scientists) should also yield the same results.** That is, the results must also be verifiable. Ultimately, these measurement results, once reported and pondered, are subject to acceptance or rejection by the scientific community as valid.

There really is no such thing as a "negative result" in a well-designed and executed experiment. If the research is conducted using the Scientific Method, the results, no matter what they are, are likely verifiable. If experimental results are reproducible and verifiable but do not agree with the theory, the results are still valid; it is the theory that is in question. This issue occasionally results in high drama in the scientific community. A notable example is the famous "ultra-violet catastrophe, discussed in some detail in Chapter 3 where the scientific community's acceptance of quantization of an atom's energy levels unfolded as high drama over a period of forty years.

Questions and Ponder Fodder for Chapter 1

1. In your own words, describe Small Space vs. Big Space.

2. Describe the scientific method.

3. Why are the two criteria, reproducibility and verifiability, important?

Chapter 2- Big-Space I

Center of the Universe

Motion

Speed of Light – The Ether

Space and Relativity

Where am I?

We begin our discussion of Big-Space with a single point in empty space; let's call that point P (Figure 2-1). So far, there is nothing there; it is just a point conjured up in our minds. Where in space is P? There is nothing in space but space. So, do we say P is in the middle of space, along the edge of space, or what? What would we mean by "edge of space" anyway? It seems that we really haven't done anything yet, and already there is a problem!

Figure 2-1: A single point in space

Now, let's put a second point in space, say Q (Figure 2-2). Again, where in space is Q? We could say that Q is near P, but where is P? Evidently, there is no reference point for empty space.

Figure 2-2: Two points in space

This is because empty space has no boundary; it's empty space! When we say that point Q is so many meters from point P, we have produced a reference point (point P) for all other points in space. That sounds great, but let's not forget that we still don't know where in space point P is. If we don't know where we are, how can we go where we want to go? What we need is a fixed reference point, like we have on our Google map; "I am here (point P), and Uber is there (point Q); here he comes!" I know I am here because I have a home address to reference against. Evidently, there is no home address for a single point in empty space.

How big is this problem for the physicist? Well, by analogy, suppose my Google home address is incomplete. For example, suppose everything is available except the state I live in. Then, as long as I look for points Q that are local (i.e., in my own state), there is no problem. But if I reach for points in other states, there is a problem, because my reference point is in an unknown state. So, in empty space, as long as we don't reach for points Q too far from P, we can keep it local and the fact that we don't know where reference point P is, is not a big problem.

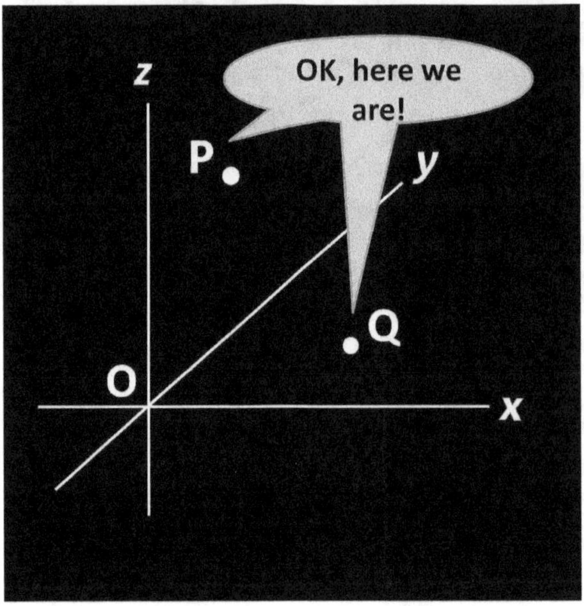

Figure 2-3: Two points referenced in space to an origin O.

In science, we often play this "keep it local" game so we can make progress. We specify the position of point Q relative to point P using a "coordinate system" (Figure 2-3). The origin, O (x=0, y=0, z=0) of our coordinate system is our reference point, and it can be whatever we want it to be. For example, I could let the origin be the center of my favorite chair in the living room of my home, and I could call that point "my spot" (like Sheldon does). Or I could be less provincial and call the origin the center of the earth (like GPS does). Or maybe I am an astronomer, and the origin of my coordinate system is a point in the center of our Milky Way Galaxy.

And then there are cosmologists. Cosmologists tend to "think big". The cosmologist is interested in the entire universe (that's right, the whole enchilada; "all that is"). Then, where do the cosmologists put the origin of their coordinate system? If they decide to put it in the center of the universe, where is that? They are back to square-one. They have no fixed reference point and no home address. Or, as we say in science, there is no stationary reference point.

The Center of the Universe

These issues of a reference point may sound a bit trivial to someone new to science. But, far from it, they beg far more important questions. For example, we will talk later about evidence that our universe is expanding. When we think of an expanding universe, we might picture in our mind the image in Figure 2-4a, which shows stars moving away from a center point, the original singularity.

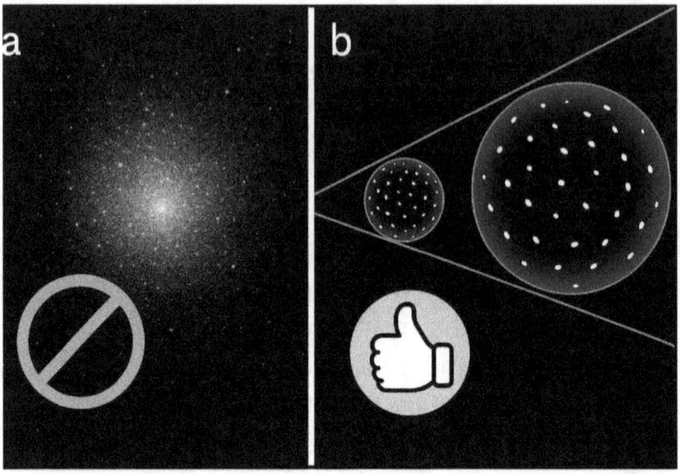

Figure 2-4: If we are told that the universe is expanding, we might imagine a bunch of points all diverging from a single point, as depicted in 2-4a. While this is conceptually accurate, it is not what we actually see. This is because we are part of that expansion; we are seeing it from within. In this perspective, all points (stars, galaxies, even the atoms in our body) move away from all other points as shown in 2.4b. In this perspective, any knowledge of where it all began (i.e., that coveted stationary reference point), is lost. (i.e., that coveted stationary reference point), is lost.

In fact, what we observe is an expansion featuring all points, stars, galaxies, etc. moving away from all other points Figure 4b. All the celestial bodies we can see from earth are moving away from us. From our perspective on planet Earth, it looks like Earth is at the center of the universe and everything is moving from us. But, if we imagine being on another planet far from earth, and we looked out into space, we would see the same thing; all other observable bodies moving away from us, leading us to conclude that WE are at the center of the universe. Evidently, in an expanding universe, all points move from all points and there is no center (Figure 2-4b).

Motion

We are learning that while the universe is a big place, it is not so much "space filled with stars" as it is "stars filled with space." To us, these stars appear static, like a portrait on the wall.

Acceleration is strictly defined in physics as a time-rate-of-change of velocity. That is, if a body moves at a constant velocity, then it is not accelerating. If its speed changes from 50 mph to 60 mph in a few seconds, then the body is accelerating. Additionally, if a body is moving at a constant speed but changes its direction while moving at that constant speed, we also say it is accelerating. So, acceleration can be a change in speed of a body, or a change of its direction of motion.

In physics, we represent acceleration using a ***vector****. It can be thought of as an arrow pointing in the direction of the acceleration with a length that is proportional to the magnitude of the acceleration. So, if I am on a merry-go-round my speed may be constant, but my direction of motion is continuously changing, so I am accelerating.*

Vectors are used to represent (mathematically)a variety of physical quantities like velocity, position, momentum, etc.

Acceleration and Vectors

For thousands of years, our view of the star formations has hardly changed; nothing appears to be moving except the planets in our solar system. But in fact, the stars we see in the sky are moving at enormous velocities; they are just so distant that their movement is not apparent. The universe is a very dynamic place! So, let's consider what the motion of particles and points in space means. If we want to understand the role of space in the functioning of the universe, we must consider the meaning of moving points, i.e., motion.

Let's revisit our two points in space (P and Q; back to Figure 2-3. But this time, let's suppose that point Q is moving away from point P at some known speed, say, 10 miles per hour. Without a reference point, all we can know is that the speed of point Q relative to point P is 10 mph. We must now ask, "Is point P moving from point Q, or is point Q moving from point P? Or is it some combination of both?"

We all have experienced the feeling while on a train slowly pulling out of the station next to another train; for a moment, we cannot tell whether we are moving forward or the train next to us is moving backward. One way to know who is moving is if we feel a jerking motion on our train; then we know WE are moving, and we know in what direction. Another way is to look at some distant object, like a building, that we know is not moving and observe whether we are moving relative to the building. If we observe that we are not moving relative to the building, then we conclude that the train next to us is moving and we are not. In this scenario, the building represents a fixed or "stationary" reference frame, and it is pivotal in resolving our dilemma of relative motion. Without a stationary reference frame, the only other way for us to know who is moving is if we feel something. The jerking motion we feel is the result of our bodies being accelerated as our train goes into motion from a full stop. Once the train gets moving a couple of miles per hour and is slowly coasting out of the station at a constant speed, we again feel nothing, and we can again convince ourselves that the other train is moving backward.

So, evidently, there is a real difference between a reference frame that is moving at constant speed and direction and one that is changing speed or direction (i.e., accelerating). In Physics, we call a reference frame that is moving at a constant velocity an "inertial" frame. If our reference frame is somehow changing speed or changing direction, we say that the reference frame is "non-inertial." For example, our Earth is a non-inertial frame of reference; Earth rotates 360 degrees every day, and it moves in a large circle around the sun. If we watch the night sky all night, we will see the stars rotating around the North Star, completing one revolution each day. But the stars are not moving in circles; we are!

Another simple example is a merry-go-round. When we are on the merry-go-round, we see the world going past us even though we are not ourselves moving relative to the merry-go-round. As we watch the world pass by, we feel something pushing us away from our seat. What we feel is the tendency of all moving things to continue in their current direction. It's Newton's first law; " A body in motion tends to stay in motion; a body at rest tends to stay at rest." As we sit on the merry-go-round, our direction is continuously changing. That actually means that we are accelerating; the merry-go-round is a non-inertial reference frame.

It happens that, as far as we can tell, that elusive stationary reference frame we discussed does not exist in our universe. In this simple claim lies the basis of at least 100 years of effort, discovery, disappointment, and wonder that we now call the golden age of Physics. In what follows, we will describe some of the events that occurred and how they impacted our view of the world we live in; let me warn you, it may be a bit of a bumpy ride!

The Speed of Light and the Ether

We are beginning to see that empty space is more complex than it appears. Actually, it is far more complex than what we have seen so far. Even today, the scientific community struggles to understand empty space. Far from being trivial, these problems of finding a stationary reference point to make a measurement are the basis of Einstein's theories of relativity, and as we will see, they are an important part of current efforts to combine our understanding of empty space with what we already understand about nature. It happens that the properties of light as it travels through empty space are strongly linked to the properties of empty space itself. But that's not where the strangeness ends. Once we include what are called "quantum effects" (we will get there!), things get very complicated and very strange; "curiouser and more curiouser!".

Light and EM (Electromagnetic) Waves

In this section we will extend our discussion from abstract concepts of an imaginary point in empty space to the idea of light passing through empty space. So, let us explore this concept by telling a story.

We all have seen what happens when we drop a stone into a pool of water; waves emanate from the point where the stone disturbed the water. These waves carry away the energy that the stone imparted to the medium, the water (See Figure 2-5). Without water, the medium, the waves could not exist. Most of this is intuitive to us; we see it every time we jump into a swimming pool or take a bath. But there are some real mysteries when it comes to wave motion.

We know that radio waves, also called electromagnetic or EM waves, exist; they bring us cable TV, cellular phone calls, GPS, radio broadcasts, and the like to us every day. EM waves also occur naturally and come to us as sunlight and radio waves from all the stars in the sky; they always have been, and they always will. But, unlike water waves, EM waves travel in empty space, in a vacuum. With EM waves, there is no direct evidence that there is a medium for these waves to travel in. How can that be? Somehow, energy from a disturbance, like the stone discussed earlier, is imparted to the vacuum, i.e., empty space, and the EM waves carry away that energy, supported only by empty space! If we stop for a moment to ponder this, we must conclude that it is very strange!

Figure 2-5: (a) Water waves, supported by the medium, which is the water. (b) Radio waves, supported by nothing but empty space, radiate like a water wave. How can this be?

In the nineteenth century, between 1860 and 1871, the physicist and mathematician Clerk Maxwell thought so. But rather than messing around with concepts of empty space, he imagined the idea of a "field" that could somehow exist in empty space. This field carries energy as though it were traveling through a medium.

James Clerk Maxwell *was a Scottish-borne mathematician who applied formal mathematical methods to some of the most vexing questions of how nature works. Most notably, Maxwell succeeded in unifying the physical phenomena of electricity, magnetism and light into one theory (now called the theory of electromagnetism), showing that all three are just different manifestations of the same physical phenomena. His theory includes a set of equations, called "Maxwell's equations". The set describes electromagnetism so accurately that they are still used today, 150 years after their publication.*

Maxwell was born into a middle-class family in Edinburgh Scotland. A scientist with a strong religious faith, Maxwell is considered by the scientific community to be one of the greatest physicists who ever lived.

James Clerk Maxwell (1831 – 1879)

Using what was then known about EM waves, along with mathematical properties of what is called a "vector space," he developed a beautiful theory of electromagnetism, one that explains almost all of what we observe in our everyday life.

It was an astounding success, the product of true genius!

But the question remained: How can a wave travel through empty space? A lot of really smart people pondered this question for many years, and one idea that emerged was the concept of the "ether" (or "aether" if you are British). It was called the "stationary luminiferous aether," but that is a mouthful. The ether was imagined to be an undetectable, stationary, that is, not moving; there's that pesky notion of motion again! but real fabric that pervades all empty space, like the tub of water in the previous example. It was imagined that the ether interacted with energy that perturbed, and in the process, it somehow supported the transport of EM waves.

Well, it is convenient to claim that the ether is undetectable, but most physicists, especially if they are experimental physicists, get edgy when someone claims the existence of something that cannot be detected. So, immediately, scientists began to conjure up ways to detect the "undetectable." It did not take long. Scientists reasoned that if the ether exists, it would have one measurable property: it has to do with the speed of light, and it has to do with the concepts of relative motion that we discussed earlier.

So, let's look at Figure 2-6 to understand the role the ether would play in the propagation of a light wave. Imagine the Earth on its path around the sun. Relative to the sun, the Earth's heading changes with the seasons. So, relative to an observer on Earth, the ether, even if it is stationary, looks like a "wind" blowing past Earth. We can compare Earth's motion through the ether to a motorboat speeding on a calm lake. As we sit in the boat, we see a cork in the water float backward past us at the speed of the motorboat. Looking perpendicular to the motion of the boat, i.e., out one of the sides, the cork and the water appear to be flowing past us, from front to back, but neither towards nor away from the side of the boat. In like manner, an object like the Earth, traveling through the ether, will pass through it as a boat passes through water. As we see in Figure 2-6, depending on the direction of motion of the Earth, the ether will act like a wind whose direction and speed depend on the direction and speed of the Earth itself.

Now, let's imagine that the same observer on Earth is sending a beam of light outward from Earth, through the ether, into space. If light is a wave, like a water wave, and the ether is the medium that carries those light waves, then that light beam will travel through the medium at some speed, let's call that speed "c" relative to the ether, just like the water wave that travels through the water medium. If light travels at a speed c relative to the ether, and if the ether wind is blowing toward us on Earth at a speed v, like a headwind, then the speed of light measured relative to an observer on Earth will be (c + v). Conversely, if the ether wind is a tailwind, then the observed speed of light, relative to the observer of Earth will be (c - v). This is all just like the boat on water; the measured speed of the wave depends on the speed of the observer.

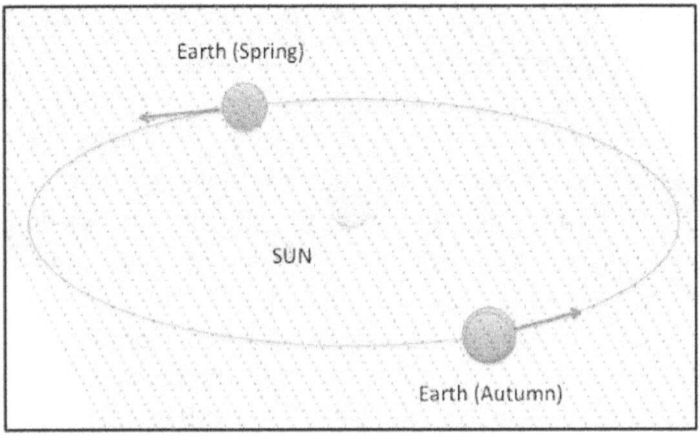

Figure 2-6: As Earth orbits the sun, the ether appears as a "wind", changing speed and direction with time to an observer on Earth

So, this is all a little tedious but very straightforward and intuitive; light moves at a speed c relative to the ether, and depending on the relative speed of the person measuring the speed, the measured value of the speed of light could be less than or greater than c. This is very tidy and clear. Scientists reasoned that if we could measure the speed of light in different directions at different times of the year, then we could map out the morphology of the ether itself, the invisible "undetectable" wind!

Albert A. Michelson (1852-1931) was a professor of experimental physics at Case School of Applied Science (Cleveland OH). Specializing in optics, Michelson was awarded the Nobel prize in 1907 for his work measuring the speed of light.

Edward W. Morley (1838-1923) born in Newark, NJ in 1838. He was a chemistry professor at Western Reserve College in Cleveland, OH. Before that, he attended Andover Theological Seminary in 1861-1864 and had a preaching appointment in a small parish in Ohio. From 1864-1868. On or about 1868, Morley began his chemistry career, culminating with his position as professor of chemistry in Western Reserve College in Cleveland, OH. He is famous for his extremely precise and accurate measurement of the atomic weight of oxygen.

Michelson and Morley

The Measurements of the Speed of Light

In 1887, using these ideas and knowledge of the earth's motion through the heavens, a scientist named Albert Michelson devised a clever way to measure c in different perpendicular directions at the same time. Michelson, a professor of Physics at Case School of Applied Science in Cleveland, Ohio, was an experimentalist. He was aware of interferometry, an optical technique used to measure the flatness of surfaces to extreme precision, less than the wavelength of light. Using this concept, he devised a new kind of interferometer, now called the Michelson interferometer, that made it possible to measure the speed of light in two perpendicular directions at the same time and compare the two results. The technique was not simple, but it would make it possible to measure the speed of light to very high precision in many different directions.

Michelson was very skilled in the lab, but his mathematical background was somewhat limited. In his early attempts at measurement, he made some mathematical errors and was unable to reconcile his measurements with his theoretical expectations. His first publications contained these mathematical errors, and the scientific community quickly alerted him to this. So, Michelson recruited a colleague named Edward Morley, a professor of chemistry at Western Reserve University, to team up with him on the project. Morley's mathematical skills were much better than Michelson's, so it was a good match. The experiments are to this day attributed to both Michelson and Morley. Notably, Michelson received the Nobel Prize for this work, but Morley did not. A short explanation of this is that even scientists are people, and high drama often accompanies scientific breakthroughs.

The experiment was not easy. The slightest movement from vibration, such as horse or carriage traffic, thunder, or ambient temperature drift, was disruptive enough to destroy the measurement. So, they set up their experimental apparatus in the concrete basement of the Adelbert Dormitory on the campus of what is now known as Case Western Reserve College in Cleveland, Ohio. It took more than two years of intense labor to set it up and get it operational. It was so intense that Michelson suffered a nervous breakdown in 1885. During his recovery period, Morley carried the baton to set up the experiment.

T. Hendrik Lorentz was a Dutch physicist who devoted much of his career investigating the nature of light and electromagnetic phenomena, especially the behavior of light in moving reference frames. Lorentz' work strongly influenced Einstein in his formulation of relativity and the theory of gravity. In fact, Lorentz was one of the first to advocate for the concept of "covariance".

Hendrik Anton Lorentz (1853 – 1928)

During a six-month period in 1887, Michelson and Morley executed a series of measurements of the speed of light in various directions at different times of the year. The measurements were very accurate and reproducible, but the results were unexpected and very surprising. In fact, the results were so unexpected that no one believed them at first (not even Michelson)! The results revealed that when it is measured, the speed of light is always the same relative to the observer, no matter whether the observer is moving, standing still, rotating, or standing on his head!

The reaction of the scientific community was initially disbelief; "Something is really wrong with the experiment; this cannot be!" If a spaceship moving at nearly the speed of light, say 2/3 of c, projects a beam of light forward, at speed c, then a stationary observer must see the light moving at speed 2/3 c + c = 5/3 c. But intuition fails us; the light is measured with a speed c no matter what!

What is a Light-Year?

A common mistaken assumption is that a "light-year" is a measure of time, like a month or year. Actually, a light-year is a unit of length. It sounds like time, but not so. One light-year is the distance a light beam travels in one year. In one year, light travels about six trillion miles. It is a little staggering to realize that the closest star to us is five light years, (or 30 trillion miles) away.

The edge of our observable universe is fourteen billion light years from us....that's six-hundred-trillion-trillion miles!!

A little diddy for the mathematically curious; in relativistic physics, we let the speed of light equal the number 1, so that rather having unit of speed (i.e., meters/second), it is assumed to be dimensionless. When this is done, time and space are equivalent. When we speak of time, we speak of distance and visa-versa.

The concept of light having a completely constant speed is counter-intuitive and it was not easily accepted. As time passed, other scientists did experiments to measure c. These results consistently confirmed Michelson's and Morley's reported results. They all confirmed the unbelievable result that "the speed of light is the speed of light, 2.99 x 108 m/sec". Measurements continue to this day; much more accurate but all confirming the unbelievable! In his "Special Theory of Relativity", Einstein asserted that not only is the speed of light fixed in nature, but also nothing in nature can ever exceed the speed of light.

Richard Feynman, one of the greatest physicists of the twentieth century, once said in a lecture at CalTech:

"If a theory disagrees with experiment, it's WRONG. In that simple statement is the key to science. It doesn't make any difference how beautiful your guess is, it doesn't matter how smart you are who made the guess, or what his name is... If it disagrees with experiment, it's wrong. That's all there is to it."

The experiments of Michelson and Morley unexpectedly closed the door on the beautifully intuitive, self-evident ether theory. Evidently, the idea, that the "luminiferous aether" pervades all empty space, and that it provides the stationary reference frame needed to make absolute measurement, is beautiful; and it is wrong! The results of Michelson and Morley's experiments left the physics community without an explanation. It took the cognitive genius of Albert Einstein to parse these confusing data into a coherent explanation of nature's behavior. It turns out that instead of the luminiferous aether being fixed in space, what is fixed in nature is the speed of light. No matter what direction light is moving, no matter whether the observer is moving, no matter what, the speed of light is always the same relative to the observer: 386,000 miles per second, 300,000,000 meters per second, or 1,800,000,000,000 Furlongs per Fortnight! At that time, there was no theory that could accommodate this new experimental reality. A new theory was needed.

Space and Relativity

The findings of Michelson and Morley triggered a revolution in physics. The period between 1880 and 1905, and then on through to approximately 1930, became the most pivotal and disruptive period in the history of science so far! Not only were the established physical theories at odds with experimental reality, but, perhaps for the first time in science history, our intuitive views of nature were proving to be inadequate.

Michelson's and Morley's measurement of the speed of light established experimentally that the speed of light is a "constant of nature." That is, its value (which is 300 Million meters per second, or in scientific notation, 3 x 108 m/s) never changes, regardless of the state or conditions of any observer, no matter how fast or what direction that observer is moving or even whether the observer is accelerating. It forced the scientific community to conclude that our understanding of what "empty space" must be incomplete.

Einstein showed in his Special Theory of Relativity that for light to have these special features of constancy, we must abandon our intuitive ideas of empty space and replace them with a new concept of space called "space-time".

This means that, in order to accommodate nature's insistence on a constant speed of light, the "fabric of space" must somehow be altered or "distorted" when relative speeds approach the speed of light. Also, the presence of massive objects, actually, any object that has mass, will distort space-time. This is one of the premises of Einstein's theory of gravity, earlier called his General Theory of Relativity.

This all sounds like the answers have been found, but actually, these "answers" raise even more questions. For example, if space, or space-time, can be "distorted" by gravitation, then must it not have a structure to begin with? This question leads us back to the postulation of an ether, which we have shown cannot exist. So we are left with a paradox; somehow space has a fabric that is not stationary, but it can support electromagnetic or gravitational fields. How such fields might interact with the fabric of space remains a mystery! But when we mathematically represent the concept of field interactions with space, the results fit the data with extraordinary accuracy!

Simultaneity and Light

Because the speed of light is finite, events that might be

correlated with each other may or may not appear so. The philosophical concept of synchronicity is not as intuitive as we might think. Events occurring at one location may be simultaneous to one observer but disconnected in time for some other observer. As we will see, signals marking the beginning and end of events are not transported instantaneously from one place to another but instead require some time to accommodate the speed of light. For example, let us consider the behavior of thunder and lightning in our everyday lives.

We all have had the experience of a lightning storm of seeing a flash of lightning and then sometime later hearing the crash of the thunder. We might notice that for every four or five seconds (actually, 4.7 seconds) of delay, the lightning strike is about another mile away. We might say to ourselves, "The storm is two miles away, because the delay was about ten seconds." We say that because we assume that the light from the lightning flash was almost instantaneously transmitted to us from the source, while the sound traveled more slowly than the light, its arrival was delayed.

But it is a little different when long enough distances are involved. Suppose out in space, let's say a hundred light years away from Earth, a star explodes in a great flash. Unlike the case of lightning, the flash of light does not come to us instantaneously. Because light travels at a rate of 386,000 miles per second, darned fast, but still not instantaneous, there is a time delay between the actual event and our seeing the flash at Earth. But wait; how do we know the starting point for the delay if there is not a "starting time"? For us to know when something occurred, we must know where the event was when it occurred. Conversely, our knowledge of the distance of an event depends on when it occurred. There is an ambiguity between space and time. Because light does not travel instantaneously, time and space are inextricably tied together. The set of "coordinates" that used to occupy the three dimensions we know now occupies the three dimensions of space plus a fourth "dimension," which is time. We call this combination "space-time," or, if you like to name things, "Minkowski Space."

Two Guys and a Spacecraft

To illustrate the concept of speed-of-light induced simultaneity, we consider the case of two separate observers, A and B, moving at different speeds while observing two events occurring "simultaneously".

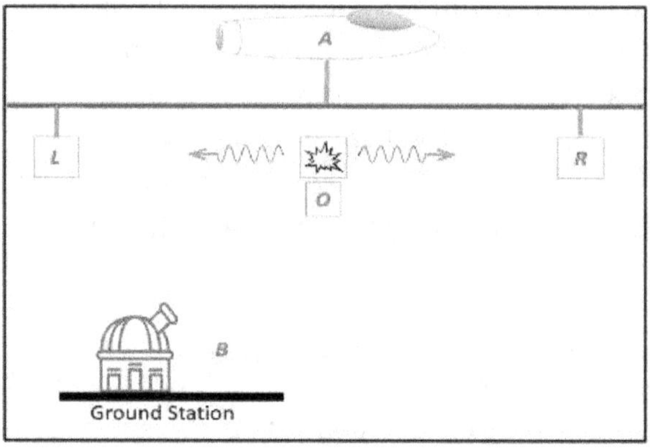

Figure 2-7: In the reference frame of Observer, A, L and R are equidistant from point O on the spacecraft, so the explosions at L and R occur at the same time; i.e., the explosions at L and R occur simultaneously.

Referring to Figure 2-7, we imagine observer A to be in a spacecraft moving from left to right, while observer B is in the ground station, which is stationary relative to the ground. Observer A in the spacecraft is traveling from left to right at a speed v relative to the ground station B, as shown. Connected to spacecraft A is a long rigid rod, which moves as A does. That is, the rod and the spacecraft move as one.

At time t = 0, a pulse of light is emitted from point O on the spacecraft. This pulse moves in both directions of the bar, toward points L and R, which are equidistant from point O and also moving with the spacecraft. When these light pulses reach points L or R, they each trigger a light sensor, which immediately sets off another small explosion to announce the arrival of the light beam.

What does observer A see?

Since the bar holding the explosive devices and sensors is connected to and moving with spacecraft A, the observer A in the spacecraft sees each of the two pulses moving the same distance at the same speed c, the speed of light. Therefore, observer A sees the two pulses a L and R occurring simultaneously.

What does observer B see?

Observer B is not connected to the spacecraft but is on the ground station (See Figure 2-8). From his perspective at the ground station, spacecraft A is moving from left to right at speed v. He observes the spacecraft/rigid rod assembly flying by from left to right at speed v.

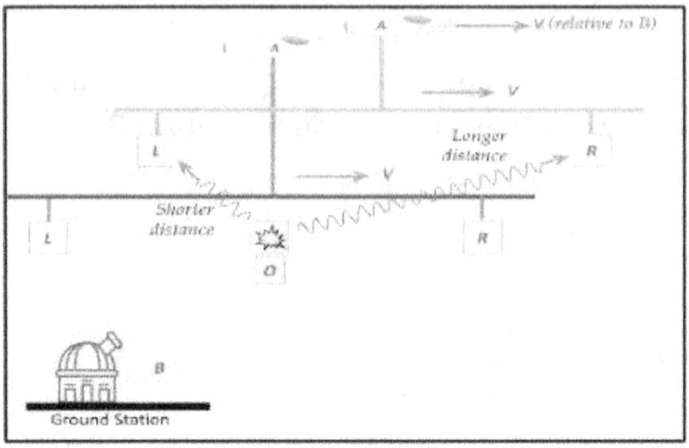

Figure 2-8: Observer B, who is in the reference frame of the ground station, sees the spaceship carrying observer A along with the rigid rod carrying the explosive devices at O as well as Land R, all moving at speed v. As the light flashes from O move, points L and R also move, making the travel distances from O to L different from O to R. In the reference frame of Observer B, L and R are not equidistant from point O on the spacecraft, so the explosions at L and R do not occur simultaneously.

When the explosion at point O occurs, the two light pulses move away from point O, one toward L and the other toward R. However, as these light pulses move toward points L and R, these points L and R, being connected to each other via the rigid rod and the spacecraft, are seen by observer B as moving from left to right at speed v. This means that the light pulse moving toward point L from O will have a shorter net distance to travel because the point L is moving toward point O with speed v. Meanwhile the pulse of light moving from O to R will have a longer distance to travel because point R is moving away from where point O was at the time of the initial explosion at O.

Therefore, the two light pulses which originated at point O simultaneously, will arrive at L and R at different times;

that is, they are seen by observer B as not being simultaneous.

Observers A and B, who are moving at different relative speeds, disagree with each other on the simultaneity of the events at L and R. This disagreement is not resolvable; it is a fundamental ambiguity of measurement, seeded by the finite nature and constancy of the speed of light.

There are philosophical implications to this “fourth dimension.” It is not always possible for us to know what is happening someplace else at this very moment because what’s happening there is too far away to know now. For example, suppose that our sun suddenly explodes; it will take about eight minutes for us to get the bad news. Is it reasonable to say, “right now the sun is exploding but we won’t realize it for eight minutes”, when we have no way of knowing this until it hits us? The idea that events some distance away are happening now, but we have a way to see them is not the reality of our universe as we know it today. Any two events, separated by space and time, are not immediately connected; the idea of “action at a distance” is evidently not tenable in our world, or as will be seen in Chapter 4, perhaps a few pieces of the puzzle are still missing.

When we see a lightning flash and hear the thunder later, there is no problem in this regard because the lightning flash marks a starting point to count, and the thunder crash marks the end-point. But we have no time reference to do the same thing when the light flashes from a star heads toward us because nothing moves faster than the speed of light; yes, I know, Star Trek, etc., etc., but really, it looks like c is the limit!

So, because the speed of light is finite, the concept of simultaneity comes into question. Two events that appear to be simultaneous in one reference frame are not in another. In fact, the whole idea of something happening in another place “right now” is called into question. “Right now” is entangled with “right here” in our world.

Questions and Ponder Fodder for Chapter 2

1. Is there a stationary reference point for our universe? How do you know?

2. What is an “inertial” frame of reference? Compare this to an “non-inertial reference” frame.

3. What is Newton’s first law?

4. How is acceleration defined in physics and how is it represented?

5. Who was James Clerk Maxwell and why is he regarded as an important physicist?

6. Why was Michelson and Morley important to science? What was their contribution?

7. What is the speed of light? Does it ever change?

8. What is a light year?

Chapter 3- Small Space I

Quantum Behavior

Quantum Beam Splitters

Quantum Diffraction

The Dual Nature of Light

Models of the Atom

"I think I can safely say that nobody understands quantum mechanics."

Richard Feynman, 1965

In Chapter 2, we explored the properties of space by looking outward to the edge of the known universe. That's a really big space, and from here on, that's what we will call; "Big Space".

Big-Space and Small-Space

Empty space is both big and small. When we consider what's happening between large objects like planets, stars or galaxies, space is Big. We have seen that in Big-Space, position and motion are important, giving rise to theories of Relativity and Gravity.

In Small-Space, interactions between particles of mass become important. When these elementary particles get close enough, the space between them somehow become part of the particles themselves. We call these kinds of behavior "quantum-like".

Size of an atom = 10^{-10} *meters*

Size of a Quark = 10^{-22} *meters*

Size of a Galaxy = 10^{+21} *meters*

Size of the observable universe = 10^{+27} *meters*

Size of everyday life = 1 meter

But empty space need not be large to be mysterious. Small space, like the space between two atoms or two sub-atomic particles, is also interesting and quite mysterious. Most physicists, if they are being honest with themselves, will admit that Small-Space is very strange and that so far, we really do not understand it. In this chapter, we will explore small space and discover that somehow, small space percolates matter out of NOTHING - what? And can annihilate matter into NOTHING - what? Evidently, small space, when it is small enough, does NOT need to conserve energy, momentum, charge conservation, or obey a host of other common physical laws; instead, somehow, small space creates and annihilates particles "at will" out of nothing and into nothing. Like a shrewd banker, Small-Space somehow borrows mass, energy, or momentum, uses it to make or destroy particles of mass, and pays back the debt before anyone is the wiser! We can use laws we observe, like the Uncertainty Principle, to explain these events, but no one really understands how these things happen or why they do; we only know, through measurement, that they do. Evidently, empty space, i.e., Small Space, is not so empty! Modern-day physics has become an intense search for the contents of the vacuum!

It can be very difficult to understand or accept the notion that small space really IS the basis of the world in which we live. Overzealous scientists do not just make this stuff up for the sake of telling a good story; well, OK, once in a shameful while, they do. This really is what our world is like!

In this chapter, we will explore some of these strange features of Small Space. We will learn that, as far as we can tell, space - ALL space, big and small is granularized or quantized, like a screen door rather than a solid pane of glass. We will learn that dynamic variables like energy, momentum, and spin are also quantized. We will learn that the process of discovering and verifying this odd behavior of nature took at least forty years of intense effort by brilliant scientists before the quantum was accepted as part of nature. We will also learn that nature has more up her sleeve than the quantum. Sometimes, light behaves like particles, like little baseballs shot out of a cannon, and other times like waves. Light can spread out or diffract as though it is everywhere at the same time. This feature of light is called the "dual nature of light." As if that is not enough, we will also learn that particles of matter, like atoms or electrons, which carry electrical currents that run our computers and light our homes, also have a similar dual nature. Sometimes, depending on the physical circumstances at play, particles behave like waves, and sometimes waves behave like particles.

Discovery of the Quantum World

The discovery of the quantum lies at the heart of the story of small space. It is an epic story, filled with mystery, intrigue and human behavior. In telling this story, we will not delve too deeply into details of the physics, although we will touch on some. Our purpose is to get a feel for what "doing physics" can be like.

The story begins in the second half of the nineteenth century. An exact beginning date is arguable, but it begins around 1870 and runs for several decades until approximately 1912. While the scientific community was beginning to confront the problems of Big-Space (i.e., relativity), a storm was quietly brewing in the Small-Space world of the atom. During that period, physics appeared to be tied up neatly in a bow. Electricity and magnetism were well understood, thanks to Maxwell's contributions in the 1880s. Newtonian mechanics (ca. 1700) described the motion of particles and planets alike with astounding accuracy. The states of matter were well understood through thermodynamics and the theory of heat. Lord Kelvin, a British heavy-hitter physicist of that time, was heard to utter the words "All that is left in physics is to obtain higher precision measurements of what we already know".

But there was a storm brewing; that comfortable feeling of completeness was about to be shattered. An undertow of questions was pulling the science community in the latter part of the nineteenth century. These questions were all related to experiments whose results could not be explained using the physics that was known at that time. We will talk about three of these happenings in the next few pages.

Atomic Spectra

One of these happenings was a series of experiments that revealed strange "lines" in the spectra of gaseous discharge tubes. These lines, called "discrete spectra," were first discovered in 1849 by J. B. L. Foucault and confirmed by Andres Angstrom in 1853. The lines in the spectra were inexplicable using any known physics at the time. For more than forty years after that first discovery, the science community studied these spectra before the phenomena could be adequately explained. It wasn't until Niels Bohr postulated his model of the atom as a "quantum entity" that this mystery was resolved.

Thermal Radiation and Blackbody Theory

This story unfolded over a forty-year period, in the second

half of the nineteenth century. It is the story of how science was able to finally explain why and how an object like a stone radiates as much energy away as it absorbs. More to the point, it is a story of how human beings can change their entrenched attitudes after doing everything in our power to avoid the change! This is the story of "blackbody radiation", also known as the "Ultra-Violet Catastrophe".

Photoelectric Effect

Another important story in the history of the quantum's discovery was the events connected with the photoelectric effect. By the turn of the century, it was understood that a small electric current flows from a metal when it is illuminated with light of an appropriate wavelength. Again, all the science that existed at that time could not explain this phenomenon. It was finally explained in 1905 when Albert Einstein published a journal article asserting that photo-electricity can only be explained if the illuminating light has properties of a particle more than a wave. This "corpuscular" model of light was one of the most important events in the development of quantum mechanics.

These three watershed events, spread over a forty-year period, marked the beginning of a new period called the "Golden Age of Physics." So, let's explore these events.

Evidence of the Quantum

The idea that nature prefers simple solutions to a physical problem is intuitive. For example, a grandfather clock having a pendulum of a certain length, when allowed to swing, will do so at a rate depending only on the length of the pendulum. It doesn't matter how heavy the pendulum is; all pendula with the same length swing back and forth at the same rate. If we make the pendulum slightly longer, then its period, the time it takes to swing back and forth, will also lengthen. Shortening the pendulum results in a quicker period. This is how a grandfather clock works.

But let's imagine a different world where we made the pendulum's length slightly longer, but no change was observed in the period of oscillation. Suppose that as we further lengthened the pendulum, still no change in oscillation period occurred until, suddenly, the next slight additional increase in the pendulum's length caused the pendulum's oscillation period to jump to a new value. Would that be intuitive? Most of us would answer, "No, this is not intuitive; actually, it is a little nutty!" This nutty situation is an example of "quantum mechanics."

In our everyday world, for the most part, continuous changes in a system's parameters produce continuous changes in the system. But, in the quantum world, discrete changes like this one occur all the time. If a pendulum clock could be made the size of an atom, the clock would behave in the quantized stepwise fashion described above, just because it is small! Conversely, if we started with an atomic-sized grandfather clock and gradually made it larger and larger, from atomic-sized to molecular-sized to protein-sized to transistor-sized to grain-of-sand sized, and so on, until it was "our world" sized, we would see that these quantum effects would slowly get smaller and more fine-grained until they are too small to see, but they would still be there. This smooth transition from the quantum world to our world and back is called the "correspondence principle." As far as we can tell, our world is a quantum world; it is just that, usually, it is too fine-grained to be perceived in everyday life. Amazingly enough, however, some of the things we depend on in our everyday lives, like the transistors in integrated circuits, are now so small that we cannot design cell phones and other circuits without including the quantum effects that are occurring there.

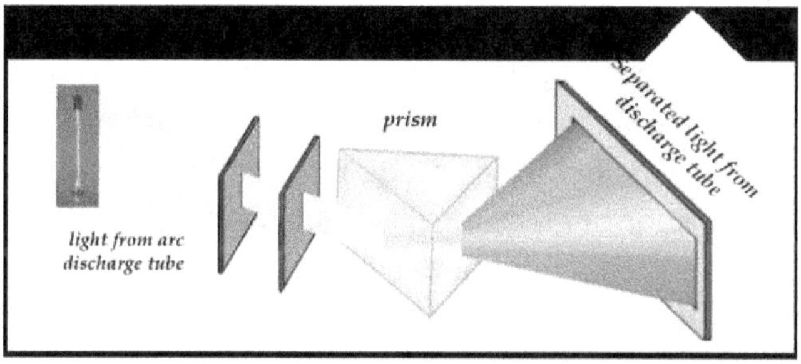

Figure 3-1: How a spectrometer works. Light from the object to be measured is guided into a prism, which separates the light's color components, showing the bluest light on the left of the screen and the reddest colors on the right side. This is similar in operation to a rainbow.

Between 1870 and 1912, the physics community was coming to grips with the idea of a quantized world. Three important quantum-effect discoveries pushed the scientific community into this new "quantum" paradigm. These were (1) the discovery of "discrete atomic spectra," (2) the "ultraviolet catastrophe," and (3) the photoelectric effect.

When scientists began to see quantum effects in the lab, their reactions were predictable; "This can't be right. What did we do wrong in our measurements? Nature isn't supposed to do this!" Now, more than a century later, the list of measurable and measured quantum effects is very long and still growing. Gradually, and over a longer period than we might expect, scientists came to accept and then embraced the idea that nature is quantized like a movie picture is made up of a series of still-motion photographs. We will talk later about a few nearly unbelievable ones.

Atomic Spectra

There is a long history of the idea that matter may be made of discrete elements called atoms. It is a natural extension of the observations made by chemists that there are a finite number of elements like hydrogen, lithium, lead, or gold, etc., that are chemically indivisible and "elemental." As early as 1808, chemist John Dalton had theorized the existence of atoms as a chemical theory. However, those concepts were "macroscopic." That is, they were only based on observations in "every-day" space scale. In 1808, no one had any idea how small an atom might be or what an atom might look like. There was no way of optically imaging objects to anything approaching Small-Space dimensions. But by the mid-nineteenth century, scientists were beginning to ponder the nature of matter at the sub-microscopic level. It was still not possible to image objects as small as an atom might be, but advances in the lab had by that time made it possible to observe a "stimulus-response" of Small-Space objects without actually seeing them. One technique that was being used successfully was the measurement of atomic spectra by means of a technique called "atomic spectroscopy." We will see how this technique gave the first measurements of the activity of the atom.

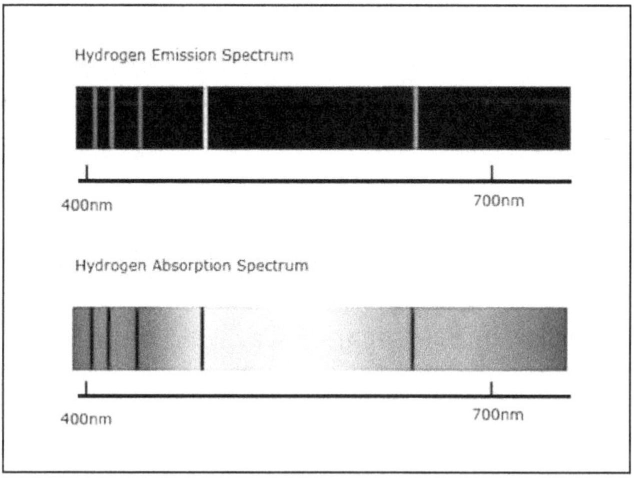

Figure 3-2: Top: Spectrum of Hydrogen vapor discharge tube. The spectrum shows discrete lines at very specific wavelengths, indicative of a quantized spectrum.

Bottom: Spectrum of the Sun, showing black Hydrogen "absorption" lines, indicating the Sun's atmosphere absorbing the Sun's own Hydrogen emission at certain discrete wavelengths.

One of the more important clues in science is what we call "atomic spectra" to ascertain what is going on in an atom. Atomic spectra are measurable light or other forms of energy that are emitted by an atom. This light can be detected in the lab, recorded, and analyzed to inform us of the goings on in an atom. An atomic spectrum can be loaded with valuable information on the atom.

By the middle of the nineteenth century (c. 1855), it had become possible to fill a glass vacuum tube with a noble gas like neon or argon and apply a very large voltage, many thousands of volts, across the tube to cause what is called “arc discharge” in the gas. The result is that the gas glows with light that is originating from the atoms of that gas. These tubes are called “discharge tubes”; they are the basis of what we all know as “neon signs.” We can capture and analyze the light from these tubes using a “spectrometer.” A spectrometer (Figure 3-1) is a measurement apparatus consisting of an optical element designed to capture and image light, a prism that separates the light into its separate wavelengths, and a means of projecting the light onto a screen.

Using a spectrometer as their measurement tool, scientists measured the spectral content of the light from many arc discharge tubes containing many elemental vapors, including hydrogen, sodium, calcium, magnesium, neon, argon, and many more. The technique was relatively easy to execute, produce, and reproduce results even in the nineteenth century. The spectra, while different for every vapor type, all were characterized by the presence of discrete “emission lines” (see Figure 3-2).

While all this was happening, scientists were using this same technique to measure the emission spectra of sunlight. That effort yielded similar results, except that different lines appeared in these spectra. Instead of being bright, isolated lines, there were dark lines situated in the middle of continuous spectra. It was later learned that these dark lines were not the emission of light but the absorption of light by the sun’s atmosphere; the sun’s atmosphere was reabsorbing the light the sun was emitting!

Detective Work

The experimental search for knowledge of atoms or other objects too small to be seen can be likened to a detective on a stakeout near a house where a suspect resides. The detective cannot enter or even approach the house. All he can do is to observe what goes in, what comes out and what changes as time passes. He must use these clues to infer what is inside the house. The physicists of that period were similarly constrained. Theorists would use their imagination to speculate what an atom might look like and how it might behave, while the experimentalists would devise clever experiments to measure what actually *goes into an atom and what actually comes out.*

Although these lines were difficult to explain when they were discovered, it was clear that they provided important clues on what was happening in small space where microscopes could not magnify enough to see. For each type of gas used in an arc discharge tube, a different set of atomic lines appeared. This spectrum of lines must be an important "signature" of that material. These observations inspired new ideas and theories on what matter was composed of at the sub-microscopic level.

Atoms as "Resonators"

Early research into small space revealed that atoms often behave as resonators. In fact, as scientific models of the atom evolved, the importance of resonance increased. Scientists like Ludwig Boltzmann and Max Planck struggled to understand how matter interacts with energy. They were consistently led back to the idea that somehow atoms are resonators; that is, they can absorb energy, store energy, and release it spontaneously or when stimulated in the right way. Depending on what a particular atom is like, how big it is, how stable it is, how it interacts with its nearest neighbors, etc., it will often resonate with its environment differently. All models of the atom, whether they are simple or complex, include this resonating feature.

When scientists first began observing these atomic

spectral lines, their initial reaction was to attribute the lines to the excitation of the atoms by the current driving the discharge tubes. In early models of the atom, the atom responds to excitation by first absorbing the energy, then storing that energy for some period, and then re-radiating that energy. This was the "resonator" concept. The idea that atoms could capture, store, and re-emit into empty space was consistent with much of the measured behavior of arc discharge tubes.

However, subsequent experimental results indicated that the atom behaved in ways that were sometimes consistent with a resonator, but at other times, its behavior was far more complex. To unravel this mystery, the only tools available were the tools of "classical physics"; these were Newton's mechanics, Clausius' thermodynamics, and Maxwell's electromagnetic theory. This was a formidable tool set, but we will see that it was not enough to treat the new mystery of the nature of the atom.

Atoms as Resonators

Most of us have had the experience of pushing a swing back and forth. It's all in the timing; we don't push until the swing seat is at the top of the swing. We keep repeating that push each time the swing seat crests at the top. Before long the swing is moving at its peak. When we match the frequency at which the swing naturally swings with the frequency at which we push the swing, the swing becomes easy to push. In physics terms, we call that action "resonance". Resonance occurs in physics all the time, and in our everyday life. I had a roommate in college who would take a recorder, the musical instrument, into the shower and play musical notes until he found the resonance frequency of the shower stall. It was the frequency at which "standing" sound waves were set up and the sound suddenly was much louder and mellower.

Each time we push, the swing goes higher. This is the principal of resonators

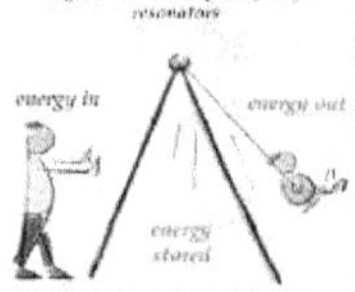

That's because the harmonics of the resonance frequency also resonate. Other examples of resonance in music are the pipe organ. The pipe length and diameter are designed to resonate at a prescribed frequency and its harmonics. Most reed instruments, the piano, the violin just to name a few act the same way. On a more exotic note, the sound we can generate by rubbing the rim of a wine glass with our finger is a resonance. It is called a "whispering gallery mode" resonance. Resonance is not always a good or beautiful thing, however. The reader might want to check out the footnote[1] to see what happened to the Tacoma Narrows Bridge in 1940 because of an unchecked structural resonance in the bridge!

The atom, it turns out, has many properties of resonators. When it absorbs energy from space, for example a light wave, it stores and re-emits energy, sometimes identical to what it absorbed and sometimes at a different wavelength.

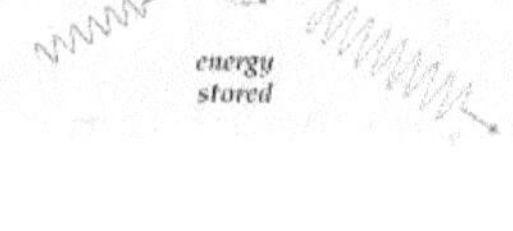

Thermal Radiation and the Ultraviolet Catastrophe

The "ultraviolet catastrophe" is a story of how the scientific community first came to accept and then embrace the concept of the quantum. It is not just about science. It is also about human nature; humankind's inherent tendency to resist change. The popular expression "If it ain't broke, don't fix it" proved itself to be costly in this instance. It delayed science's inevitable introduction of the necessary "quantum" into physics.

By the end of the nineteenth century, it was becoming clear that a breakthrough was needed to move science forward. Intuitive models of the atom, like the resonator concept just described or a simple planetary model, were not adequate to account for the measured behavior of atoms. This was a special time in physics history when common sense or even intuition would not suffice. Rather, as we shall see, a radically different vision of nature was called for, one that our five senses fail to capture. Even Max Planck, the main author of these radically new concepts, didn't really think they could be right. He fought against his own ideas for decades as he slowly went about permanently changing the course of physics. These new concepts revolved around the concept of the "Quantum," the idea that nature prefers empty space to be laid out as a grid system rather than a continuum. It isn't just space that is evidently quantized. Other physical quantities like energy, momentum and spin also are quantized by nature; and as usual, we have figured out how, but we don't know why!

The reader must keep in mind while reading this section that the characters in this story were brilliant thinkers with minds more agile than most of us have ever encountered. For these people, the idea of thinking "outside the box" was irrelevant; for them, there is no box. Problems were solved using a deep familiarity of nature's nature and mathematics, the language of science.

But even these great minds were subject to nature's credo of homeostasis. The required scientific paradigm shift was so great that even the most brilliant minds of the time resisted. The community of science was and still is just that, a community. As with all communities, diverse opinions coexist, expectations vary, and consensus must be somehow achieved before any significant change can occur. In the case of the quantum, this change was a bit painful; the process took more than forty years, but it happened, and once it happened, the change was permanent, like the clicking of a ratchet. We might begin to say that the story of the quantum was an epic story of biblical proportion!

The Story

The story begins on or about the second half of the nineteenth century; around 1870. As discussed earlier, science was "quiet". It looked like the scientific community understood nature just about as well as it was ever going to. It appeared that only a few loose ends were left to tidy up. One of those loose ends was the question of whether matter is continuous or discrete. For example, does a lump of clay exist as a continuum of matter, as it appears to, or does it consist of many smaller particles called atoms like a handful of fine-grained sand, for example. If matter is made up of individual atoms, then it must not be continuous. Or, if matter is distributed discretely, then between these atoms must be our old friend, empty space (i.e., Small Space). What holds matter together in this case? Is there some sort of "field" as described by Maxwell; (see chapter 2), somehow filling empty space and serving as the "glue" that holds the atoms in their places? Or is it some mysterious property of space itself that is mediating the behavior of atoms? So, if an atomic model of matter is correct, then maybe the science community didn't really understand matter as well as they believed. But the science of discrete matter and the mathematical tools needed to treat such a problem were also limited. All there really was at that time was mechanics of continuous matter, electricity and magnetism and thermodynamics. Of these tools, thermodynamics was most suited though not really sufficient.

Ludwig Boltzmann (1844-1906)

Ludwig Boltzmann was born in Vienna, Austria in 1844. He was a physicist and mathematician who spent the better part of his career formulating a theory now known as the Kinetic Theory of Gases. By the 1870s it was clear that matter was made up of atoms. But precious little was known about them. Boltzmann devoted his entire scientific career to developing a "statistical" theory of heat that could account for an atomic view of matter. His theory served as the stepping-stone to quantum mechanics.

Boltzmann's Work on the Atom as a Resonator

Ludwig Boltzmann (1844 - 1906), a physicist and mathematician, was keenly interested in the nature of thermal radiation. He wanted to figure out the exact relationship between the behavior of atoms and the observed phenomenon called thermal radiation. This curiosity, on Boltzmann's part, marked the beginning of the first substantive discovery of the quantum.

In our everyday lives, we all are familiar with thermal radiation, even though we may not realize it. When we gaze into a fireplace and watch an ember glowing red, yellow, or even blue, we are seeing thermal radiation. Any mass body, like a stone or a lump of clay, will emit light or thermal radiation. However, at the same time, that mass body will also absorb radiation. This is how that rock or lump of clay interacts with its environment. How it interacts depends strongly on the temperature of the environment in relation to the rock or lump of clay. If the rock is warmer than the environment, then it will emit more radiation than it absorbs as its temperature cools to the environment's temperature. Conversely, if the rock is cooler than the environment, it will emit more radiation than it absorbs as its temperature approaches that of the environment. If the rock's temperature is exactly equal to the environment's temperature, then the rock will emit exactly as much radiation energy as it absorbs. This condition, i.e., when the temperature of the object is the same as the environment's temperature, is called thermal equilibrium. So, if the rock is in thermal equilibrium with its environment, it will emit as much radiation energy as it absorbs.

Boltzmann's interest in the thermal radiation problem was driven by his desire to understand atomic matter, how atoms interact with each other and their environment to produce what is experimentally observed and measured. His approach to the problem was grounded in his knowledge of mechanics and thermodynamics. Boltzmann reasoned that if matter is made up of atoms, then those atoms would likely interact with each other through collisions, much like billiard balls on a pool table, but on a much grander scale. There would be many trillions of collisions occurring every second, with all these atoms exchanging energy and momentum back and forth. In fact, the scale of the problem is vastly more than that; no one really knew at that time how complex the problem really was. Even if Newton's laws could describe each billiard ball interaction, there could be upwards of 1023 (a trillion trillion!) equations to manage and solve simultaneously. This problem is so daunting that it was virtually insoluble using the tools of the day. A different approach was needed.

Boltzmann well understood the complexity of this problem. His solution to the problem implemented a clever game plan. Rather than even attempting to calculate the position and momentum of each atom in such a massive ensemble, he would calculate average quantities and estimate the likelihood that these averages were correct. All he would assume about the atoms was that they were tiny hard spheres moving around and colliding with each other like billiard balls. He assumed that each particle had its own energy by virtue of its jostling motion but that this motion was totally random and immeasurable within the material. Boltzmann set out to calculate the average energy of these particles so he could ascertain the total internal energy of a lump of mass.

Boltzmann's Equipartition Theorem

As a mathematician, Boltzmann understood the remarkable but not-so-intuitive features of randomness in nature. As intuitive creatures, many of us would nod our heads and agree with the claim that if two events are random, then two hundred events must be even more random. But in nature, very often, the opposite is true. When random events occur with larger and larger elements in a sample, the computed result becomes more and more predictable! This is why, for example, election results can be projected early after the polls close. As the number of votes increases, the prediction becomes more accurate. Statisticians can even calculate the probability that the prediction is correct. So, they know when they can project a winner with, say, 90% confidence once the sample size is sufficiently large. These are not arbitrary predictions. They are based on the way nature actually works. Again, as in earlier discussions, we now know that this is how nature behaves; we just don't know why!

Boltzmann also understood that if he could characterize the randomness of the atoms in a lump of clay, he could compute extremely accurate average results. Remember, there are more than 1023 atoms in a lump of clay. Using these ideas and his knowledge of mathematical statistics, he produced a remarkable result, which he called the Equipartition Theorem. By this theorem, the average energy per atom in a sample is the same for every atom, and it is proportional to the temperature, T, of the sample. This result asserts that, although matter made from atoms is subject to tremendous randomness, the aggregate matter will consistently present the same properties when measured many times. Somehow, nature evens out these trillions of random collisions in a way that results in an orderly and predictable picture at the macroscopic level.

This is an amazing result! Without knowing anything about the makeup of the atom, or even if atoms exist, Boltzmann had succeeded in connecting the macroscopic world of matter, energy, heat, and mechanics with the, at that time, undetectable world of the atom and, amazingly, the approach greatly simplified the mathematics of the solution. Often, when a scientist manages to describe nature accurately, the explanation is simplified. Boltzmann's success in describing matter in atomistic terms opened the door to an entirely new field of research, now called "statistical mechanics," which formed the basis of what is now called "quantum mechanics."

The Equipartition Theorem was very important because, for the first time, it was possible to connect a measurable quantity, i.e., the total energy in a lump of matter. to whatever was going on at the atomic level, even though nothing was really known about the atoms. It helped scientists to use the well-established theory of heat and to push deeper into the microscopic world of matter. We will soon, however, see that this fundamental result was the source of the discrepancy between the classical world and the new "quantum" world.

Principal of Detailed Balance

It was well understood by the mid nineteenth century that a body of mass exposed to radiant energy will both absorb that energy and re-emit it as well. It does so in a way that establishes a steady-state equilibrium between the tendency of the blackbody to heat up and its attempt to temper the heating process by re-emitting radiant energy back into the environment. This process of establishing equilibrium is now referred to in science as the "Principal of Detailed Balance". All material bodies in nature interact with each other through detailed balance. Somehow, the molecules and atoms in these bodies of mass arrange their energies so that they emit and absorb radiant energy in a way that the mass-body's temperature is stably set at a value that puts it in equilibrium with its environment. If something changes such that the body of mass's ability to absorb energy (we call this the material's absorptivity), detailed balance will act to rearrange all the molecules and atoms' energies to establish a newer temperature, hotter or colder, depending on the change that took place.

Wolfgang Boltzmann first theorized the Principal of Detailed Balance by 1872. It elegantly anticipated atomic theory before the presence of atoms was established.

Isn't nature amazing?

What is Thermal Radiation?

By the middle of the nineteenth century, it was understood, through experimentation, that any sample of matter like a rock, for example, or a lump of clay, that is at a temperature T will emit, or "glow" radiation, and it will also absorb radiation. The warmer the body, the "bluer" the radiation, and the brighter it is. Also, the body emits the same amount of radiation that it absorbs. The rock is said to be in thermal equilibrium with its environment at temperature T.

From our everyday experience we know that a hot object might glow red, while even hotter, the object will begin to glow bluish. This glow is radiation emitting from the rock. The tendency to glow more bluish the hotter the object is called "Wien's Displacement Law". In scientific parlance, this glowing tendency is called "thermal radiation."

Global Warming and Detailed Balance

A good intuitive example of all this is our planet earth. Our earth is a rock that is being irradiated by energy from the sun. That energy, when absorbed by the earth, converts to heat, another form of energy, mostly in the infrared wavelength range. The earth will radiate as much energy as it absorbs by the temperature that allows equilibrium to occur. Complicated absorption and emission processes must take place through nature to maintain this equilibrium condition between earth and our solar system. This process is referred to in the scientific community as the "principle of detailed balance". Earth's atmosphere is a layer of matter that interacts with the energy coming from the sun and the energy being emitted by the earth. Through detailed balance, the amount of energy being absorbed by earth and the amount being re-emitted must remain equal. So, if the atmosphere changes in a way that more of the normally re-emitted energy cannot escape, then equilibrium is disrupted and the temperature of the earth rises enough so that detailed balance achieves equilibrium at a higher temperature; that is to say, the earth warms up. This is a much-simplified explanation of the "global warming" principle.

Our Universe as a Thermal Radiator

By the Principal of Detailed Balance, all bodies in the universe absorb and emit radiation continuously at a rate determined by the temperature of the body. In fact, there is very strong evidence that our entire universe behaves as a huge thermal radiator, or as we will describe later, a "blackbody" radiating energy as if it were a lump of mass sitting at a very cold temperature of 2.7 degrees Kelvin. That's just a few degrees above "absolute zero", the lowest temperature possible. Scientists believe that our universe started as a very hot unbelievably hot, like 10^{28} °K single point. That point, also referred to as a singularity evidently exploded and expanded into what our universe is now. It continues to expand and cool down even today.

The evidence for this Big Bang theory is the ever-presence of radio background radiation, microwave radio emission in the frequency range between 300MHz to 300 GHz. Arno Penzias and Robert Wilson, two physicists from Bell Laboratories in Holmdel, N.J. discovered it in 1964. They were doing microwave communications research when they encountered a "noise" source that was impeding their radio receiver's sensitivity. It was everywhere; a background that prevented detection of the faintest signals. After exhausting the usual possibilities using telescope instrumentation, signal dependent noise, birds, etc., they began looking for other candidates. In collaboration with Robert Dicke, cosmologist from Princeton University, they were led to the inescapable conclusion that the "microwave background" is a remnant of our early universe, created at the instant of its birth. As the universe continues to expand, the temperature will continue to cool down more towards absolute zero.

For their efforts in discovering the "cosmic microwave background", Penzias and Wilson were awarded the Nobel Prize in Physics in 1978.

Planck and the Quantum

Around 1880, another scientist named Max Planck, who was 14 years younger than Boltzmann, emerged in the scientific community. Planck had just completed his Ph.D. dissertation on thermodynamics. Like many physicists of that time, Planck was not satisfied with the prevailing theory of heat developed by Clausius. While it was accurate and self-consistent, it was a macroscopic theory. Planck, like Boltzmann and others, sought a microscopic theory that included the atomistic view of matter while still being consistent with the macroscopic picture of classical thermodynamics. The rub in this ambition was nobody knew what an atom was or even if it existed, how it worked or most importantly, how to measure atomic behavior.

Blackbody Radiators

In contrast to popular belief, the term blackbody does not refer directly to the color of a body of matter. We often hear that a black shirt or a black car will be hotter because it is black. This is not necessarily so. The ability of a material to absorb radiated energy depends on other factors like surface roughness and texture. It is possible that a diffuse white material with surface roughness or unusual surface properties can be much more absorbent than a smooth black surface. Also, the scientific idea of a blackbody means that it absorbs energy over an entire spectrum of wavelengths ranging from deep ultra-violet to far infrared. This spectrum far exceeds the spectrum of colors our eyes can perceive. So again, a rock or a car that looks black to our eyes may not be black over the full spectrum of radiation. It may not be very absorbent in the infrared region, which is often where a blackbody would absorb most of its energy.

The Measurements

We have established that bulk matter consists of discrete elements, called atoms. Boltzmann theorized that these atoms behave like "resonators", that can absorb energy, store energy and re-emit energy at the same wavelength or at different wavelengths. These processes, by Boltzmann's and Planck's reckoning, were shown to be "irreversible".

Despite these facts, Boltzmann was able to show mathematically through his Equipartition Theorem that even though the process of reaching thermal equilibrium was wildly random (or "stochastic"), the outcome was a state in which all atoms in the ensemble carried the same average energy. This triumph connected, for the first time, the mysterious world of Small-Space with the orderly macroscopic world of Big-Space that we all live in, introducing to the scientific community a new discipline called “Statistical Mechanics”.

With this success, both Planck and Boltzmann turned to the question of what happens within a bulk material like a rock, for example, when it is placed in an environment at temperature T that is different from the material’s temperature. This was an important problem at the time because classical physics treatments were predicting results that contradicted experimental observations. We must recall that experimental results drive theory, not the opposite. If a theory cannot accurately predict experimental results, it is wrong. This is why scientific measurement must always be reproducible, not only by re-doing the same experiment but also by other experiments done in different ways by other scientists. In the case of thermal radiation, it took many years, nearly forty years, of careful measurement and re-measurement to establish reproducible results. These results clearly contradicted the prevailing classical theory.

The reader must realize that the unfolding of this scientific dilemma was not one of rapid-fire results and response but rather a slow evolution of thinking about how matter behaves. Boltzmann, Planck, and many other scientists devoted their entire careers to resolving this problem. The result was a true watershed in the history of scientific discovery.

The work was both theoretical and experimental. The theoretical work involved applying known physics, mechanics, thermodynamics, and electromagnetic theory to the problem of thermal radiation to develop a formalism that could predict what ought to be measured.

There was significant experimental work as well and it was

critically important. It provided the needed reference point for assessing the efficacy of the theory. The measurements of thermal radiation were quite challenging. They covered a very wide wavelength spectrum, calling for new measurement techniques and even new materials. So, when experiments were devised that called for new measurement venues, theoreticians would go the extra mile to calculate the expected results in the new experimental venue. And so, it was in the case of thermal radiation.

To make the best measurements possible, materials called "blackbodies" were chosen for the experiments. In scientific parlance, the term "blackbody" refers to any material that can absorb all impingent energy across a wide wavelength spectrum. While it is true that all mass bodies will both emit and absorb thermal radiation, most of them are imperfect. Some may reflect much of the radiation rather than absorb it. Others may absorb or emit thermal radiation over some wavelength regions but not others. A few materials, like bismuth black, a very diffuse black paint can absorb and emit over the full wavelength spectrum including visible light, UV (ultra-violet), deep-UV, IR (infra-red) and far-IR. These unusual blackbody materials were chosen for use in the thermal radiation experiments because the experimental variations could be better managed in the lab.

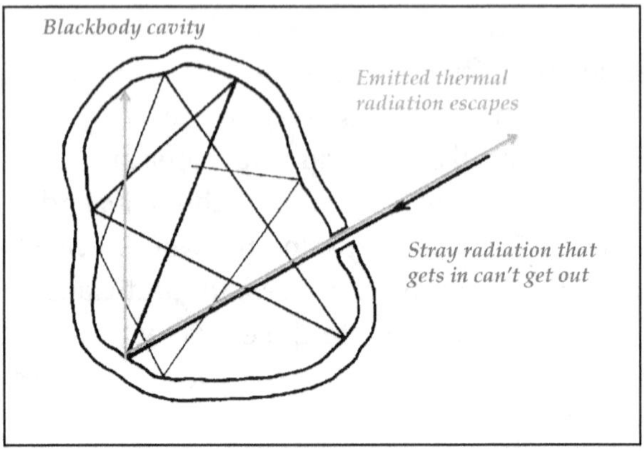

Figure 3-3

Experimentalists made another big change. Rather than deal with a rock or a lump of clay in the lab, they turned the rock inside-out. That is, a blackbody cavity was made, as shown in Figure 3-3. It was a shell coated on the inside with blackbody material and fitted with a single small access hole. The cavity was then heated to the desired temperature. Thermal radiation from the blackbody filled the cavity, and a tiny fraction escaped through the hole. The escaping radiation was almost purely thermal radiation since the only other contributor could have been radiation that somehow got in through the tiny hole, bounced around in the cavity, and found its way back out of the hole, a very unlikely scenario.

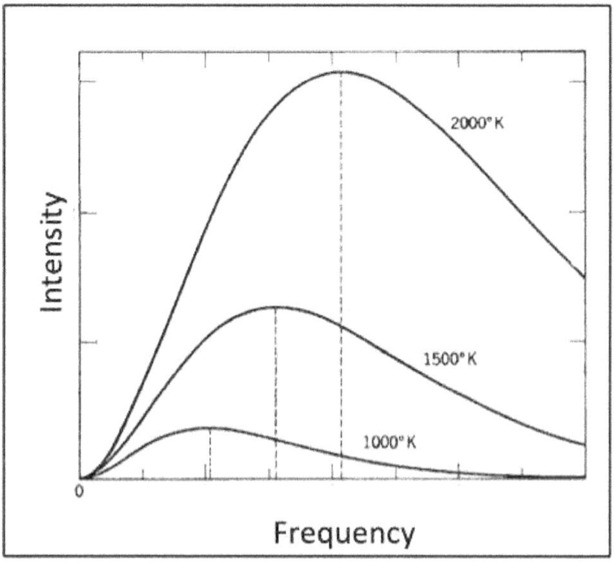

Figure 3-4: Spectral curves of thermal radiation from a blackbody radiator at different temperatures.

In the experiments, carried out by several scientists, the thermal radiation spectrum was measured over a wide range of frequencies from deep UV to far IR. The results, shown in Figure 3-4, reveal a few features.

For each temperature of the blackbody, there is a wide spectrum of frequencies or wavelengths emitted from the blackbody. This spectrum ranges from the deep ultraviolet (UV) to far infrared (IR). As temperature is increased, the total energy in the blackbody, called the radiancy, increases very rapidly. This is called Wien's displacement law.

For each temperature, the radiancy curve reaches a maximum and then tapers off as frequency is further increased. As the temperature of the blackbody is increased, the maximum emission frequency increases proportionately. This is called Stefan's Law.

In all cases, the measured energy emitted from a blackbody became diminishingly small at high frequencies, that is, at shorter, UV wavelengths.

This third feature was the most profound of all; not because it was unreasonable, because it wasn't, but because it directly contradicted the prevailing theory of thermal radiation by Rayleigh and Jeans. The reader must recall one of the first axioms of science; if a theory cannot reproduce measured results, it is wrong, no matter how elegant it is, or who authored it.

Figure 3-5 clearly shows the contradiction. The solid black line on the plot shows a typical measured blackbody radiancy spectrum, while the dashed black line shows the prediction of Rayleigh and Jeans. The measured spectrum should by this time be familiar to the reader. The dashed line representing the theory of Rayleigh and Jeans, predicts a monotonic increase in radiancy with frequency.

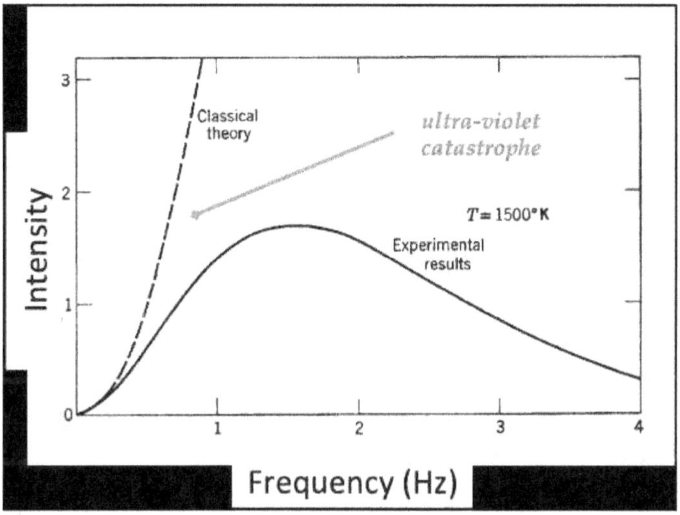

Figure 3-5

Well, something was wrong, and it was not at all clear what the problem was. Theoreticians like Boltzmann and Planck realized that whatever the problem was, it was very fundamental. Some experimentalists publicly underscored the severity of the discrepancy by pointing out that the theory predicts that at room temperature, a rock should glow with a bluish color.

Planck conjectured that there must be something wrong with the way their theory perceived how resonators (atoms) interacted with each other. After all, precious little was known about atoms in 1900, so their first assumptions could be wrong. He and Boltzmann had come to the conclusion that the means by which energy is emitted or absorbed, called vibrational modes, was a bit arbitrary, but the prevailing attitude among theorists was that although the modes of vibration and the energy carried by each mode could be counted by differing means, the end result should be basically the same. This assumption turned out to be false. Not only was the means of counting important, but the size of each counting bin also mattered. Planck began to realize that this could only be true if the type of counting and the bin sizing had some real physical significance. So, Planck began a process of testing various counting and binning methods, comparing results with the measured results described earlier. The task was tedious and arduous, but Planck noticed that if he made the bin sizes just the right size, he could fit his calculated spectrum to experimental results with impressive precision.

Planck's summation method amounted to an inbred requirement that the radiant energy term must be proportional to the frequency of the thermal radiation. This summation had to provide just the right mathematical space between terms. However, the summation requirement was defined by nature, not just mathematical preference. Planck found that at the best fit to experimental measurements, a multiplying constant, now called Planck's constant, was required to balance his radiancy calculations.

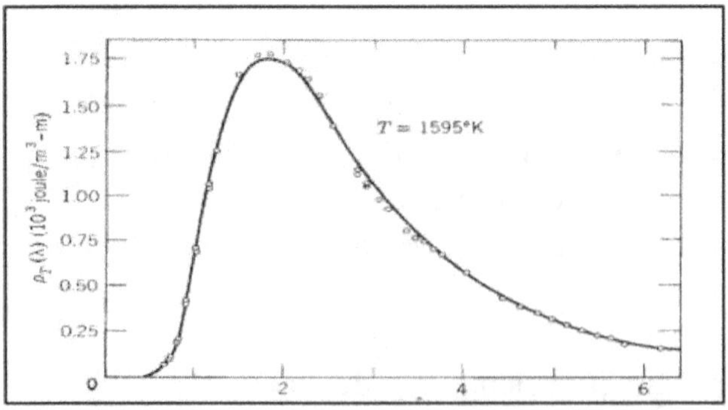

Figure 3-6: Planck's energy density prediction (solid line)

compared to the experimental results (circles) for the energy density of a blackbody. The data were reported by Coblentz in 1916 and apply to a temperature of 1595 °K.

The implications of this success were far-reaching. Summing up mathematical terms according to a single summation convention defined by nature implies that energy is quantized, not just mathematically, but physically. Some energy values are "allowed," while others are not. The result, called "Planck's distribution," led to a near-perfect fit to experimental data, as shown in Figure 3-5.

As we will see later, quantum matter appears not only in energy but also in momentum, spin, and space itself. But, as far as we can tell, time is not quantized.

Planck discovered all this not through divine inspiration or even by superior intellect but more likely from dogged tenacity at working the mathematics until he found an open path he could pursue. Once he found that path, the real physics of his solution began to unfold. Planck began to realize that his mathematical manipulations had led him to the idea that, for some reason, nature does not allow energy to exist as a continuum but rather as very specific discrete entities. Up until that time, there was no reason, other than this "Ultraviolet catastrophe", to believe that such a quantization of nature existed. The concept was not just new, but it was counter-intuitive, and to this day, much of quantum mechanics is still counter-intuitive. Speaking metaphorically, the energy in a blackbody does not fill a continuum, but rather, it looks more like a wire screen. Only at the vertices where the wires cross is energy allowed by nature, and as we have seen before, Planck was beginning to understand how nature works but understanding why was a different matter.

It is important for the reader to ponder this before moving on. What clues can nature have given to encourage a scientist to even imagine that energy could be quantized? Was there anything in our everyday lives in the late nineteenth century that would even suggest quantization? After all, at that point in time, the only hint of the quantum was the observation of line spectra in arc discharge tubes. But as we discussed earlier, these lines were being dismissed as evidence of invisible resonators. This was a partially accurate interpretation of the material.

So, as hindsight might predict, the physics community pushed forward for forty years, roughly from ca. 1870 to 1910, making two steps forward and one step backward, accumulating facts and discarding incorrect assumptions, until the reality of the quantum was too compelling to ignore. As the fictional sleuth Sherlock Holmes said, "Once you eliminate the impossible, whatever remains, no matter how improbable, must be the truth."

For his insights, efforts, and tenacity, Max Planck was awarded the Nobel Prize in Physics in 1918.

It is time to interject, at this point, a comment on the methods taken by Boltzmann and Planck. The reader might be struck with the impression that a lot of this work looks like aimless meandering. In fact, this is partially true. The physics community was, at that point in time, concurrently coming to the reluctant conclusion that on the one hand, their total knowledge of physics was inadequate to the task, but on the other hand, it was all they had to work with. A breakthrough was needed! In the real world of science, answers of this type were not to be found in the back of the book. Real science research is often very ragged and messy; it can appear to be "aimless." But it is the way the human brain engages any form of the unknown and solves problems. Scientific breakthrough is often messy business, with ironical twists and turns that make the journey especially interesting. In short, science is not so much about answers as it is about questions.

The Photo-Electric Effect

Among the handful of items to be tidied up before victory could be declared was a well-known phenomenon called the "photo-electric effect". Originally discovered by Heinrich Hertz in 1887, the photoelectric effect is the physical phenomenon that occurs when light is shined upon a metal. It was found that illuminating a metal surface with light caused a small electrical current to flow from the metal.

Figure 3-7: Classical view of the photoelectric effect. In this model, photocurrent is liberated if the light is "bright" enough.

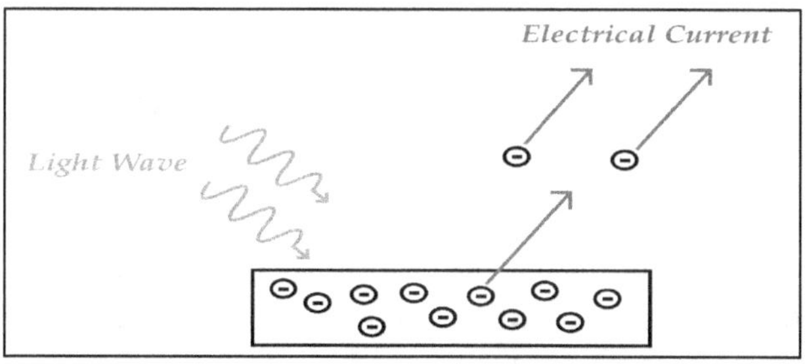

What Was Expected

The prevailing explanation for the photoelectric effect at that time is shown in Figure 3-7. Light was viewed as a wave, as we discussed earlier, that carries energy through space. Illuminating the surface of the metal with the light wave imparts this energy to the atoms in the metal. If the metal absorbs enough energy from the light wave, charged particles can get knocked from the metal into the free space outside the metal. If that space has a voltage applied to it, a current will flow. This is the photoelectric effect. In our everyday lives, photocells are used in our cameras or garage door openers and in our automobiles' self-driving systems.

It should be intuitive that some minimum amount of energy is required to liberate charge carriers from the metal. The interface between the inside of the metal and the free space outside the metal poses an energy barrier like a wall. Without this barrier, there would not be an interface; space and matter would be indistinguishable from each other. A charged particle, like an electron, must overcome this energy barrier to escape. We call this energy barrier the "work function." When light illuminates the metal, it energizes those charged particles, helping them overcome the energy barrier and escape from the metal.

This classical model of the photoelectric effect is simple, intuitive, and easy to visualize. Based on this model, it should also be intuitive that increasing the brightness of the illumination should provide more energy for electrons to overcome the work function and escape. This would result in more photocurrent.

If no current was flowing, then increasing the brightness of the light should eventually result in a current. Even if the illumination was very slight, the energy being pumped into the metal should build up until, perhaps after a substantial delay, photoelectrons begin to be ejected from the metal.

Despite the intuitive simplicity of this model, the data disagreed with the model. A reaction one might have been was to view the experiment as a failure. However, no experiment that produces reproducible results is a failure; the data speak for themselves whether they agree or disagree with the theory. We shall see that in this case, the experiments led to the discovery that light has "multiple personalities".

What Was Observed

The photoelectric effect experiments revealed that the color or wavelength of the light used to illuminate the metal was an important factor in the production of the photocurrent. There is a threshold wavelength for photo-electricity to occur. That is, if the illumination light is "too red" (i.e., the wavelength too long), no current is produced, no matter how bright the light. On the other hand, if the wavelength of the light is "more bluish" (i.e., shorter wavelength), then current flows, no matter how dim the light. Not only that, but a dimly illuminated metal surface produces current immediately, with no time delay.

These results, at the time, referred to as the "Lenard effect," contradicted everything scientists "knew" about how light interacts with matter. For several years, many experiments attempting to verify or refute the Lenard effect were carried out. However, the experiments were quite difficult for the available technology at that time. The results of these important experiments were inconclusive.

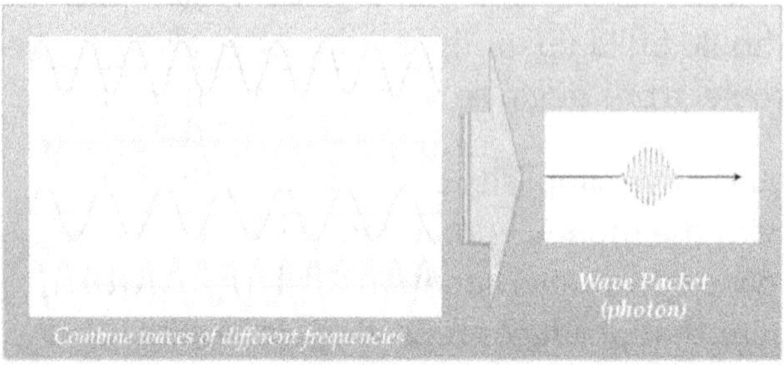

Figure 3-8: The photon is viewed as a packet of light energy having a specific wavelength (or a narrow band of wavelengths). Mathematically, a photon can be constructed as careful combination of many continuous waves as shown. This means that while the photon's energy is mostly confined to the packet, it actually extends over all space.

Meanwhile, a budding new German theoretician named Albert Einstein was postulating a new explanation to the Lenard effect. His explanation challenged the scientific communities view of light. Einstein reasoned that the nature of light was more complex than being a continuous wave that carries energy through empty space. He theorized that somehow a beam of light might be a conglomeration of particle-like wave packets. These wave packets were the combined result of many continuous light waves, forming a self-contained packet of energy. We now call these wave packets "photons"; a term coined by Frithiof Wolfers and Gilbert N. Lewis later on in 1926.

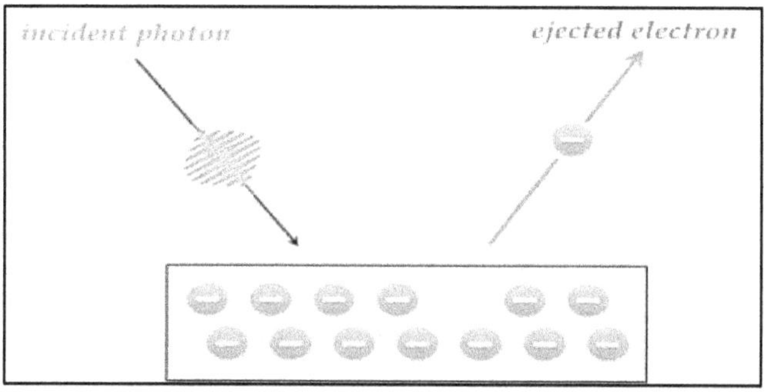

Figure 3-9

The photon, if it existed, would have some properties that are not wave-like but more particle-like. The wave packet would be made up of many waves, combining in a way that results in its energy being compressed into a small area, as shown in Figure 3-8. Mathematically, a photon can be constructed as careful combination of many continuous waves as shown. This means that while the photon's energy is mostly confined to the packet, it extends over all space. As we will see later, there are interesting implications regarding the knowledge of "where" or "when" is a photon.

Einstein reasoned that if light had wave packet properties, then an individual photon could carry enough energy alone to free an electron from the metal (see Figure 3-9). The amount of energy of each photon would have to be proportional to the wave frequency of the photon or, inversely proportional to the photon's wavelength. So, with the photon model, the measured existence of the wavelength threshold is a natural outcome. Also, since a single photon can liberate an electron from the metal, there is no "build-up" of power needed to produce a current, so no time delay should occur. The time delay was a problem of the classical photoelectric effect since no such time delay was observed in the lab. However, in Einstein's model, no time delay is expected.

Atoms

Up until this point, we have been looking at small space from the point of view of early pioneers of the quantum. To be able to understand modern physics' picture of small space, we must follow the subsequent developments that led to our current picture of the atom and its constituents.

The notion that matter is made from atoms, or some smallest indivisible entity, is not new. The Greek philosopher Democritus first proposed this notion in the fifth century B.C. Before then, several civilizations, including the Greeks, modeled our world to be made up of the four or five basic elements: earth, air, fire, and water or energy. Apart from the notion of indivisible building blocks, no detail was available; however, how could it be? There was no way to image or measure sizes anywhere as small as an atom.

Imaging the Atom

By the eighteenth century, technologies became available for scientists to image objects as small as a few microns, in particular, the achromatic objective. This is still much larger than an atom but small enough to observe evidence of the atom's existence. In 1827, a botanist named Robert Brown made an important discovery. He observed, under a microscope, rapid chaotic motion of pollen grains suspended in a still liquid. It was as though the pollen grains were being knocked about without cause by some invisible force.

This behavior, now called "Brownian Motion," was theorized by Albert Einstein to be caused by energetic molecules or atoms randomly colliding with the pollen grains. Einstein published two papers on his ideas on Brownian Motion in 1905; the same year, he reported his theory of the particle nature of light (i.e., the photoelectric effect). In the same year he published his theory of special relativity, and his theory of the equivalence of matter and energy ($E=mc^2$). It is noteworthy to mention that while Einstein was awarded a Nobel Prize for his reported work on the photoelectric effect, it is clear that any of these five publications could have earned him the Nobel Prize. It was a very fruitful year for Einstein. In fact, it was the culmination of many years of mental labor.

Getting to a scientific consensus that there is a smallest entity called an atom was one thing; figuring out what to do about it is an entirely different question. Atoms are so small that they could not be directly viewed with a microscope, like a cell or a virus can. Instead, the scientist had to infer what is going on in the atom by "perturbing and observing". But there are limits on what we can do. Like the detective on a stake-out, we cannot interview the deliveryman; but we can watch what he does after he comes out of the house. It is in this spirit that the scientist designs and carries out experiments to measure the "immeasurable".

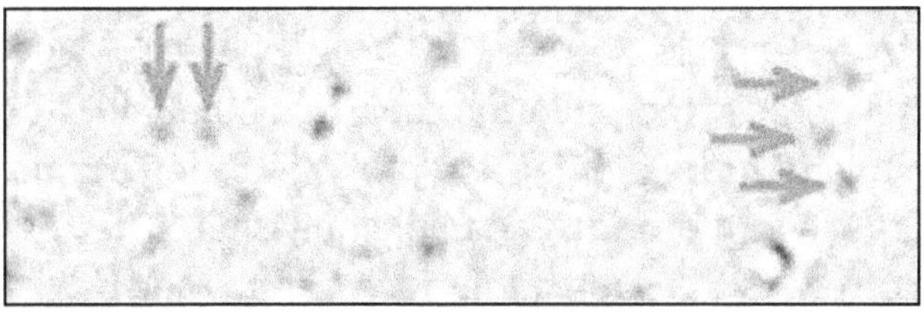

Figure 3-10: Electron microscope image of single hydrogen atoms.

In present-day technology, it is almost feasible to directly image the atom, which is 10^{-10} meters small. However, imaging of any structure in the atom is limited by technology, and ultimately by basic science principals. Figure 3-10 shows the best raw image of a hydrogen atom currently possible without resorting to computer aided processing. Presence of the atoms is evident, but structural details are blurred out. This image may look unimpressive to the reader, but we must remember that we are looking at an object that is one hundred thousand times smaller than a human hair!

Imaging the atom may seem very ambitious, but we now know that the atom is not the indivisible entity that its name implies. Much smaller particles called nucleons reside within the atom along with electrons. Imaging these entities by themselves is much more challenging; possibly hopeless, even in the future. It isn't clear where it ends; it looks like nucleons are made from smaller entities called Quarks, whose dimensions are much smaller yet. The atom, as small as it is, is mostly our old friend empty space. It looks like using pure imaging means to capture a picture of the atom and its constituents will not do; the particles are simply too small to "see".

Thompson's "Plum Pudding" Model of the Atom

By 1895, there was some scientific consensus that all matter is made of more elementary particles called atoms, from the Greek word atomus meaning indivisible. The exact structure was the subject of much debate and speculation. Experimental evidence implied that the atom was very small, electrically neutral, and somehow it could configure its structure to make all the elements that were known to exist. The actual size of the atom could not be ascertained until the early twentieth century, when it became possible to produce collisions of energetic beams of particles with gold foils in the lab. These experiments implied that the atom was much smaller than was previously assumed. Subsequently, these experiments revealed more information about the structure of the atom.

J. J. Thompson proposed one model favored by the scientific community at the time. In his model, the atom is a positively charged sphere of material filled with tiny particles called electrons that carry a negative charge (see Figure 3-11). He called his model the "plum pudding model." In a series of experiments with cathode ray tubes, like the old picture tubes in TVs, in 1897, J. J. Thompson discovered the electron and measured its mass and charge. He also demonstrated that the electron is 1,000 times smaller than the atom itself. He received the 1906 Nobel Prize for his discoveries.

Figure 3-11: The Thompson "plum pudding" model of the atom. In his model, the entire atom is filled with positively charged material, while the inside is dotted with negatively charged electrons, resulting in a "net-neutral atom.

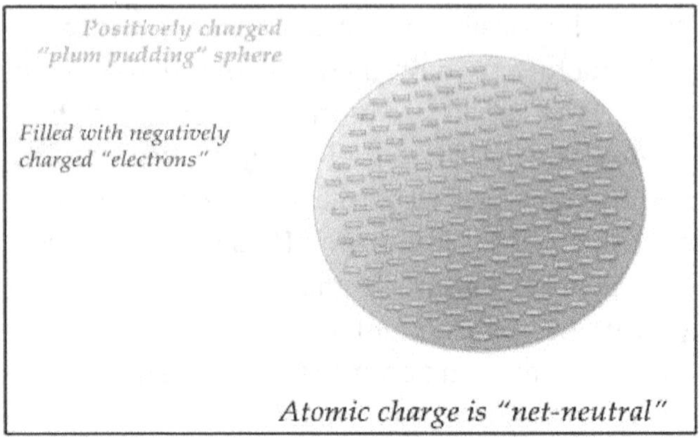

Although Thompson's model was a thoughtful one and led to his discovery of the electron, it was found experimentally to be incorrect. Atoms that were bombarded by heavy particles often recoiled straight back as though the atom was much more massive than Thompson's model implied. The only reasonable explanation was that the atom was structured more like a solar system; the heavy nucleus was at the center of the atom, much lighter electrons orbited the nucleus, and a lot of our old friends had empty space between.

Rutherford's Planetary Model of the Atom

One of J. J. Thompson's students was Earnest Rutherford. Rutherford conducted a series of lab experiments in which energetic atoms of Helium, also known as alpha particles, were used to bombard a thin foil of gold. The alpha particles were "scattered" in many directions, and these data were recorded. The surprising result was that a small fraction of these collisions resulted in the bombarding alpha particles recoiling straight back from where they came as if they had collided with a brick wall. This result was difficult to explain using Thompson's "plum pudding" model. In his model, the material is evenly distributed through the atom, with no dense material to cause the small fraction of violent collisions that were measured. The data implied that somewhere inside the atom was a massive but small particle that was causing the "back-scatter" effect.

In response to this new data, Rutherford proposed a new model of the atom that he called the "planetary model" (see Figure 3-12). The atom is modeled after our solar system, with a center object, like our sun, called the nucleus and planetary objects called electrons orbiting the nucleus. The nucleus is assumed to have a positive electrical charge, and each electron orbiting the nucleus contains a negative charge. The atom is assumed to be "net neutral" in its total charge, so the positive charge of the nucleus is equal to the sum of all the negative charges of the electrons.

Figure 3-12: The planetary model of the atom, while simple and intuitive, is not tenable in describing the measured behavior of atoms in our world.

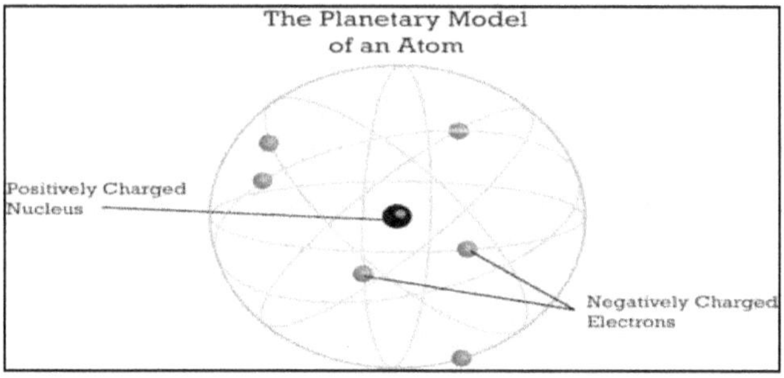

Rutherford's model of the atom has some elegant features since it resembles the solar system, and if it were only correct, it would be a very gratifying model, but it is not correct. There are some differences between a solar system and an atom with protons, neutrons, and electrons that cannot be resolved. The biggest problem with the planetary model is that unlike our solar system, which operates under the gravitational attraction of masses, the particles making up the atom carry charge; their motion is driven by electromagnetic forces, not gravitational ones. It turns out that electrically charged particles that are moving in curved paths, like a circle around a nucleus, lose their energy during such motion. So, an electron in orbit around a nucleus would rapidly radiate its energy away while orbiting the nucleus and spiral out of its orbit into the nucleus, causing the early demise of the atom. This is not observed to happen in an atom. If it did, matter would spontaneously disintegrate. Instead, matter in our universe is, for the most part, as stable as our universe. Most atoms are stable against decay for billions of years. Of course, heavier atoms like uranium or plutonium are less stable, but that is a very separate discussion.

Bohr's Quantum Model

We learned earlier that the discovery in the 1850s of discrete line spectra in arc discharge lamps was surprising, and it could not be explained by any known physics of the time. In the decades that followed, as efforts increased to understand the structure of the atom, this gap between theory and experiment widened. The physics community was confident that, with enough effort, the phenomena could be explained using known physics. More and more sophisticated mathematical techniques were applied to the problem to fill the gap but to no avail. By 1910, it was clear that Rutherford's planetary model of the atom could not accommodate the measurements of emission by atoms of discrete energy. Something was wrong somewhere.

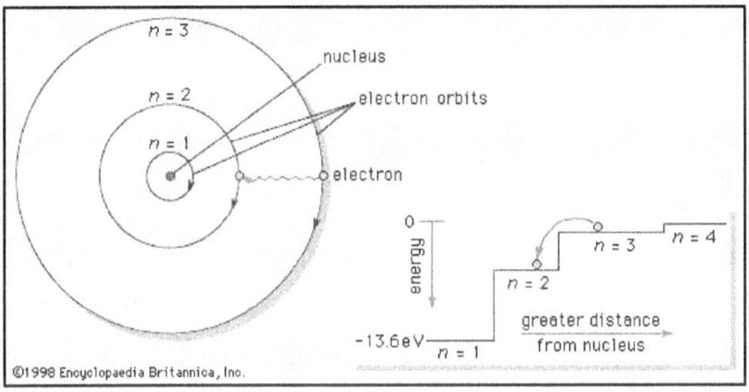

Figure 3-13: The Bohr model of the atom

Finally, out of these mathematical attempts emerged a much bolder or, more desperate perhaps, model of the atom. In 1913, Niels Bohr proposed a radical modification of Rutherford's planetary model: part physics, part "gut feel," and part extrapolation logic. The concurrent emergence of quantum effects, like the photoelectric effect, influenced the thinking of scientists like Bohr. Perhaps the corpuscular nature of light and matter applied to observations of atomic spectra as well.

Bohr's theory asserted that there were orbits for the electrons in an atom but that only certain orbits were "allowed" (see Figure 3-13). Bohr's math implied that these allowed orbits had wave-like features. However, unlike Rutherford's model, Bohr's orbital mathematics more resembled "standing waves" like the ones we might set up in a slinky. With these features in his orbital mechanics, he could have a planetary-like model with standing waves, which are stationary and, therefore, do not radiate energy away.

Niels Bohr

Considered by some to be the father of quantum mechanics.

Born Oct. 7, 1885

Died Nov. 18, 1962

Danish

Nobel Prize 1922 donated the gold medal to support the Finnish war effort against Hitler.

One of the Founders of CERN

The price he paid for this feature was that the orbits were "quantized"; only certain orbits were possible. But it was worth the price because discrete orbitals provided a "natural" way to account for the discrete lines measured in atomic spectra. It also fits well with Einstein's quantized theory of the photoelectric effect.

To express this in physics terms, Bohr asserted that one of the important parameters of such an atom, called "angular momentum," can only exist as integer multiples of a number or a constant we call "Plank's Constant." The reader may recall Planck's constant from the Ultraviolet Catastrophe. Bohr showed that "quantizing" the orbital angular momentum of an atom resulted in "preferred" or "quantized" orbits for the electrons orbiting the nucleus. The position and energy of these orbitals agreed approximately with the spectral lines observed in arc discharge tubes.

It was clear that Bohr was onto something, even though the experiment was imperfect. Before long, others were improving upon Bohr's model. The simple model of Bohr led to the electron shell theory. It was becoming clear that Small-Space was somehow "quantized."

There was no known way to justify the selection of certain

orbitals over others in the atom; it appeared arbitrary. But evidently, nature had figured out how to do that. All the atomic spectra that could be measured are presented as "line spectra," as we discussed in this chapter. The Bohr model of the atom could not, and still cannot, be derived from "ab initio" or from first principles. It represented a bridging of what he knew from experimental measurement and what "looked right" from a mathematical perspective. This was one of the first times in the history of Physics when scientific theory could not flow continuously from what we could see experimentally and what our brains imagined was happening. We will often see this theme recurring in twentieth-century physics. More and more, intuitive physical insight was being supplanted by mathematical insight and what we would have to call "mathematical intuition."

Questions for Ponder – Fodder from Chapter 3

1. How is measurement important in understanding Small-Space?

2. What is "Quantum-like" behavior? Give one example of quantum-like behavior.

3. Mention three discoveries that contributed, over many years, to the Golden Age of Physics?

4. What is the Correspondence Principal?

5. What is an atomic spectrum?

6. What is a spectrometer used for?

7. Who was Ludwig Boltzmann and what was his contribution to modern Physics?

8. What is a resonator? Give an example of from your everyday life.

9. How is thermal radiation an example of the principal of Detailed Balance? How does it relate to global warming?

10. What did Penzias and Wilson discover, that led to their award of the 1978 Nobel Prize in Physics?

11. Define what a black body thermal radiator is.

12. What was Max Planck's contribution to science?

13. Toward understanding the photoelectric effect, what is the "work function"?

14. What is a photon and how did its definition evolve?

15. In 1905, Einstein published five journal articles, any one of which was Nobel Prize quality. Which one brought the Prize?

16. Can the atom be imaged by modern technology?

17. What was J.J. Thompson's "plum pudding" model of the atom?

Chapter 4- Big Space II

Doppler Effect in Space

The Expanding Universe

The Big Bang Theory

Theory of Relativity

Cosmic Microwave Background

The Discovery of Gravitational Wave

"Everything should be made as simple as possible, but not simpler." A. Einstein

Today, our knowledge of Big-Space is very dynamic, but it is far from complete. Early on, in the evolution of Big Space, the scientific understanding was driven by the experimental results of Michelson and Morley. Their discovery that the speed of light is constant, independent of all references, triggered imaginative thinking in the scientific community and illuminated our understanding of big space. It set the stage for establishing the theory of relativity, which includes both special theory and general theory.

In Chapter 3, we learned that one consequence of Michelson's and Morley's discovery is the compelling conclusion that Space must not have an ether. In response to this discovery, much of the scientific community began an all-out search for an explanation of how space can support wave propagation without an ether. There must be a medium of some kind to support the transfer of energy carried by the waves. But what is it, and where is it? This unanswered question triggered new research for some of the best scientific minds of the time. All through the twentieth century, and continuing into present time, several stunning discoveries of the nature of Space (both Big-Space and Small-Space) have occurred. Some of these research topics were related to existing field theories, such as electromagnetism and fluid dynamics.

Others focused on big space by defining its physical properties.

Doppler Effect

Doppler effect occurs in our everyday lives. For example, when a car or train passes by while its horn is blowing, the pitch of the horn will be higher than normal as the train approaches, than when the horn is receding from an observer. If an object moves towards the light source, the wavelength of the reflected light will become compressed, emitting bluer light. If the object is moving away from the light source, the wavelength is lengthened, or made red. Astronomers use this Doppler technique to measure the orbit diameter and the orbit time for distant planets. It is also used to measure the speed and distance of distant objects moving away from our spatial positioning space in cosmological expansion.

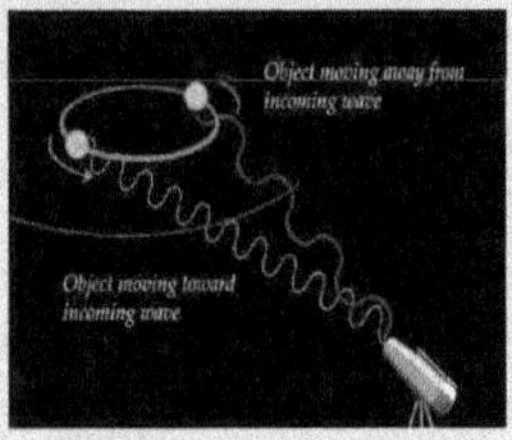

Notable among these discoveries were:

The Expanding Universe: Experimental observations of the presence of highly distant objects in deep space triggered exploration. This activity resulted in the discovery that virtually all objects are moving away from each other; i.e., evidently the universe is expanding. Hubble, through astronomical measurements, of what is now called, Gravitational Red-Shift, reported his observation that all objects in our universe are moving away from each other at speeds that scale linearly with respect to their distance from each other. This is called Hubble's Law.

Big Bang Theory: In response to Hubble's Law, a new theory on the origin of the universe was formed. According to this model, the universe was violently borne out of the explosion of a point singularity. The theory claims that the universe has been expanding ever since that time. It was originated by the Belgian Catholic Priest

& cosmologist Georges Lemaitre (1931). Along with several collaborators, Lemaitre developed his theory. Today, the Big Bang Theory is still among the most favored explanation of our origin.

Cosmic Microwave Background: The unanticipated discovery, by Penzias and Wilson, of strong experimental evidence for the existence of the Big Bang as origin of the universe.

Discovery of Gravitational Waves: Barry Karish, Kip Thorne and Rainer Weiss discovered in 2015 that gravitational energy can be transported over a "medium" called spacetime.

The Theory of Relativity

While all these discoveries and developments were taking place in the physics world, scientists like Albert Einstein, Hermann Minkowski and Jules Henri Poincaré were sorting out the structure and meaning of Relativity. The two theories of Relativity in their completeness are beyond the scope of this book, even when math and equations are eliminated. If the reader is interested, there is an excellent text on Relativity for individuals unfamiliar with mathematics[1]. For our purposes, we note that despite its complexity, the driving features of Relativity are expressed by the following statements of the two theories:

Special Theory:

Special Theory Principal applies to inertial frames of reference. The velocity of an electromagnetic wave in a

---

[1] Gardner, Martin, "Relativity for the Million", The Macmillan Company, 1962.

vacuum has the same value in all inertial frames of reference. (as discussed in Chapter 2) The fundamental laws of physics have the same form in all inertial frames of reference. It is impossible to perform an experiment in physics that will single out one particular frame as being at absolute rest.

General Theory:

General Principal - The laws of physics can be formulated in such a way that they are valid for any observer, however complicated his motion.

General Principal of Equivalence - No experiment performed inside an enclosed laboratory can distinguish between the effects of a gravitational field and the effects of an acceleration of the laboratory relative to the stars.

Figure 4-1: Shown in this figure are the measured movement of the oscillation frequencies of the shifts that take place when a moving distant body is moving toward or away from Earth. This "red-shift" effect enabled the discovery that our universe is expanding.

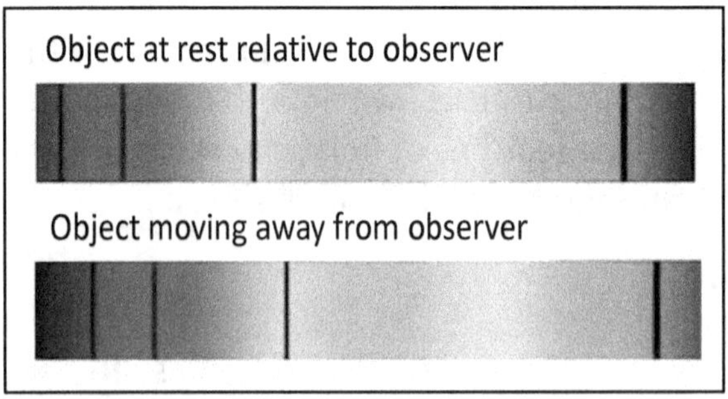

An Expanding Universe

During the period between 1912 and 1924, experimental evidence accumulated supporting the idea that our universe was expanding. This evidence came in the form of what is now referred to as "cosmological red-shifting." Astronomers who were seeking out and observing new objects in the deepest space of our universe, in particular Vesto Slipher from Lowell Observatory in Flagstaff, AZ, were looking for solar systems like our own, with planets orbiting the star. A useful tool for the work was the red-shift of the light coming from the orbiting planets. As shown in Figure 4-1, the red-shifted light is prominent when atomic lines are emitted from these far-off objects. When observing planets orbiting stars, the light would be shifted to red when the planets were moving away from the observer and blue when objects were moving toward the observer. Edwin Hubble reported his observation that all observable objects in our universe are moving away from each other. This is called Hubble's Law.

Because the objects being observed were so far away and so small-looking when viewed from a telescope, the design and construction of these apparatuses were very challenging. The primary goal of these observations was the discovery, observation, and categorization of planets orbiting the stars that were identified. The expected footprint of such an object would be alternating red-shifted/blue-shifted signals, as described in the info-box Figure 4:1. While this footprint was seen when a planet system was found, much of Big-Space was populated by objects that carried a red-shifted wavelength signature, i.e., the objects were moving away from their observer. In fact, the data implied that nearly everything in Big-Space was moving away from everything. Our universe is expanding! Now, about a century later, scientists are still convinced that our universe is expanding.

Figure 4-2: Andromeda Galaxy, discovered in 1924 by Edward Hubble et al. Such galaxies when discovered marked a new period of discovery for astronomy.

The Big Bang Theory

In the early twentieth century, it was discovered that the observable universe is expanding in all directions. This was very new knowledge, and it triggered intense interest into the cosmology of our universe. This acquired knowledge carried the mark of an "explosive event"; of a single point, called a "singularity" in mathematical terms, occurring at some time deeply in the past.

Astronomers and cosmologists were busily pondering the size and extent of the known universe.

Meanwhile, more and more mysterious new objects were being discovered in Big-Space. These objects were called “nebulae”. The word itself harkens to the French word “nebula”, coined in 1679, and the earlier Latin derivative, “nebulosus”, meaning “indistinct or vague”. These newly discovered objects were indistinct and vague. This was because of limited resolution power of telescopes of that time. What was clear, however, was that these new objects were galaxies, located far from our own “Milky Way” galaxy (Figure 4-2). They were somewhere near the edge of the observable universe, much further away than any other objects known to exist in Big-Space. Furthermore, virtually all these objects if not all of them, were found to be moving away from Earth and from each other, at very high speeds, approaching the speed of light relative to earth.

While the effort proceeded to identify these new receding objects, now referred to as Galaxies, theoreticians in the physics and astronomy communities worked to develop a model that could explain the physics of an explosion of the magnitude that implied these objects and even the closer objects are receding from us here on Earth. It was as if the entire universe was expanding! None of the existing theories addressed the question “How and why did we all get here anyway?” The theory would have to accommodate an explanation of an unbelievable event.

One explanation to this data is that the entire universe is expanding as the result of some sort of violent event. This theory came in 1931 from a Belgian Catholic Priest & cosmologist named Georges Lemaitre (Figure 4-3). Along with a few collaborators, Lemaitre developed a theory that the expansion of the universe was initiated by a point source mass “singularity” that exploded in all directions and it is still expanding.

The Big Bang Theory is still a work in progress at this time, undergoing modifications and additions as data from advanced experimental platforms like the James Webb Space Telescope, launched in 2021, the Hubble Space Telescope, launched in 1990 and others.

Figure 4-3: Priest & cosmologist Georges Lemaitre with Albert Einstein

"But wait," as the saying goes, "there's more!" It has to do with the Big Bang theory.

Recall that we briefly discussed the Big Bang theory, in the section on Thermal Radiation in Chapter 3, entitled "Our Universe as a Thermal Radiator". This is a story of an "accidental" discovery of what is now called the "Cosmic Background Radiation" (or CBR). It is arguably the most important physics discovery in the twentieth century. The unanticipated discovery in 1964, by Penzias and Wilson of strong experimental evidence for the existence of the Big Bang as origin of the universe.

Story of The Microwave Background

Figure 4-4: Horn antenna used in 1964 by Penzias and Wilson to discover the CMB. (Credit: NASA image).

In the 1960s, while debates continued over the best theoretical model to describe our universe, a most important discovery was quietly in the making, called the "Cosmic Microwave Background Radiation", (or CMBR for-short). It was discovered in 1964 that big space is filled throughout the universe with microwave radio waves of unknown origin. The team was led by Arno Penzias and Robert Wilson from Bell Laboratories in Crawford Hill, New Jersey (see Figure 4-5).

Figure 4-5: Arno Penzias and Robert Wilson, co-discoverers of the "Cosmic Microwave Background Radiation" (CMBR), which supported the Big Bang Theory.

Bell Laboratories was the research hub for the Bell Telephone Company and enjoyed a reputation as one of the most prestigious research centers in the world. Some say CMBR was accidentally discovered, but it might be more accurate to say the discovery was "very timely." It was the right time for an important discovery, and the right people were in the right place.

Penzias and Wilson were physicists who specialized in Astronomy and Cosmology. They were also experts in microwave communications systems design, since much of the astronomical signals of interest are often microwaves, and they are almost always very small signals. Both Penzias and Wilson were recognized experts in detection and amplification of exceedingly small microwave signals.

They were designing a very high-sensitivity microwave radio receiver for the Telstar communications channel between an earth-based tracking station and a balloon satellite. Telstar's goal was to become one of the world's first operating satellite communications systems. It happened that astronomical signals of interest were occurring approximately at the same frequencies (i.e., like radio stations) as the Telstar program. Both Penzias and Wilson, who were physicists, knew this, so it is not surprising that their engineering and science interests overlapped these two subjects. While their primary goal was to make Telstar a success, they were looking out of the corner of their eyes at the astronomical implications of this overlap!

The microwave receiver was designed to operate at frequencies in the 4 to 5 GHz range (~ 10 cm wavelength). Figure 4-4 shows the receiver and the horn antenna mounted on a rotating base platform. Because of its size and shape, it was lovingly called the "horn antenna." The design goal was to develop a microwave communications receiver operating in the five to six GHz range with thermal noise approaching absolute zero. This was a very ambitious design goal. Many readers might be surprised at the size of the antenna, but it is really very simple; the larger the antenna, the more microwave energy can be collected and, therefore, the smaller the detectable signal.

After more than two years of development, the receiver was assembled and finally ready for its first test run. The receiver front-end was cooled to three degrees Kelvin for maximum sensitivity. However, the first run of the microwave receiver revealed a significant noise floor nearly twice as large as expected. I am sure that this was a bit disappointing, but there is a reality of the design and implementation of complex systems such as this microwave receiver.

Seldom does everything work perfectly the first time it is

fired up. So, Penzias and Wilson were likely dismayed a bit, but it just meant that there was some engineering trouble-shooting left to be done. They needed to examine and categorize each known source of noise and then combine these elements to form a picture of the system as a noise generator and measure each suspect noise source. They looked at atmospheric radiation, an imperfection in the antenna structure, possible microwave noise from the Milky-Way Galaxy, receiver "front-end" noise, and an assortment of man-made sources of noise, including urban lights, even pigeon droppings inside the horn antenna! These efforts to account for the noise sources continued for nearly a year.

Figure 4-6: Announcement in May 21,1964 of the discovery of the 3K cosmic microwave background.

The New York Times

Signals Imply a 'Big Bang' Universe

24-Hour Dominican Truce Is Accepted by Both Sides

Penzias and Wilson, being astronomers, had been entertaining ideas that the noise source might be a real physical signal, previously not included in the list of possible "noise" sources. One such source would be a byproduct of a Big Bang universe, one that we looked at in Chapter 3. In a Big Bang universe, it all would begin with an immensley hot single point in space, mathematically referred to as a singularity, that would expand at enormous speed, cooling as it expands. This expansion and the thermal radiation that would have to follow would be everywhere. It would be isotropic, the same in all possible directions, perhaps but not necessarily homogeneous, and its frequency content would be a Planck distribution, and that distribution would be indicative of a Planck distribution with a temperature T.

Meanwhile, researchers at Princeton University in New Jersey had been considering this same possibility and were willing to share data with Penzias and Wilson. Fitting the calculated spectrum of the thermal radiation yielded a temperature of 2.7 degrees Kelvin, the temperature of the Universe!

On May 21, 1964, the New York Times front page aired the announcement that the cosmic microwave background had been discovered. For their contribution to physics, Penzias and Wilson were both awarded the Nobel prize in Physics in 1978.

The Search for and discovery of Gravitational Waves

In this section, we will give a brief account of the events that took place as a research team discovered Gravitational Waves for the first time. The effort began roughly in 1947 and led to the first identification and measurement, some sixty years later (yes, I said sixty years!) of the Gravity Waves emanating from the collision of two black holes.

Discovered in 2015, it demonstrated that gravitational

energy in the form of waves can be transported over a "medium" called spacetime. The medium itself was not detected, however.

Figure 4-7: LIGO—BIG Physics G-Wave detector. Interferometer design, 3-in-one design to accommodate the three transmission modes with a cost of > $200 M.

On September 14, 2015, at 7:50:45 US Central time, a small ripple in spacetime passed by our earth, moving at the speed of light. It was barely detectable, but it was detected, and it resounded like an earthquake throughout the scientific world. It was the first verified detection of a Gravitational Wave. It was borne from a collision of two black holes, 1.3 billion light years away and 1.3 billion years ago. It was a signal that marked the beginning of a new era in science!

Gravitational waves were predicted by Einstein's General Theory of Relativity (i.e., the Theory of Gravity). Both Einstein and Hermann Minkowski, a mathematician and research colleague of Einstein, had strong belief that Gravitational Waves existed. However, in this business there is a world of difference between "strong theoretical belief" and experimental verification or "existence-based data-driven confidence". If the Gravitational-Waves exist, the magnitude would be 10-19 times (that is, 0.0000000000000000001 times) smaller than the then current state-of-the-art in detectability. Once again, we have a situation in physics where theory predicts the existence of a physical phenomenon based on data not yet existing and that is not believed to be measurable! Despite these serious issues, a group of experimentalists put their heads down and found a way to measure the unmeasurable. We will tell the much-abridged story of the search for Gravitational Waves; a story that has continued fifty to sixty years.

Figure 4-9: Hermann Minkowsk

Early Work

In our investigation of Small-Space and Big-Space, scientists have found an important feature of Space (Big or

Small).

That is its ability to transfer energy from one place to another via waves. For example, sound waves propagate energy when the molecules that are sequentially aligned produce a domino effect in the medium, such as air or water. In January of 1957, after Einstein's death, an organization called the Gravity Research Foundation (GRF) sponsored a conference on "The Role of Gravitation in Physics" held in Chapel Hill, North Carolina. This conference hosted about forty speakers, including Richard Feynman, John Wheeler, Julian Schwinger, and other next-generation heavy hitters in quantum mechanics. For gravity wave physics, this event constituted a watershed, which made the discovery of the first evidence of kinetic behavior of inertial material.

Figure 4-8: Hermann Poincare'

One of the attendees of the conference was Joseph Weber, a physicist at the University of Maryland, in College Park, MD. Weber, who was credited for his discovery and demonstration of the first MASER (Microwave Amplification through Stimulated Emission of Radiation), became very interested in the concept of gravity waves.

The Experimental search for Gravitational Waves

From the time when both Einstein and Poincare' published their theories of gravity (c. 1915), interest in the subject grew steadily, not only in the theories but in seeking out experimental verification that either Gravity Waves are a part of nature, or that it isn't. This Scientific method process is as important as the science itself. We recall once more the words of Richard Feynman on unchecked theories, "It doesn't matter how beautiful your theory is; it doesn't matter how smart you are. If it doesn't agree with experiment, it's wrong. In that simple statement is the key to science."

Early on, these scientists began calculating the signal size they might expect to see from large events far, far away. The result of these calculations was not encouraging; the signals received at Earth would be smaller than the smallest detectable signal by a factor of 10^{+19}. Just as a reminder, 10^{+19} is 0.00000000000000000001. Needless to say, the chances of detecting a small signal looked quite grim. However, the best way to get a physicist to succeed in a seemingly hopeless problem is to inform him/her that it cannot be done.

The experimental search for gravitational waves began around 1955 in a small lab at the University of Maryland. Early in the search for Gravitational-Waves, scientists like Joseph Weber began calculating estimates of the signal size they might expect to see from large Gravity

Waves, or "G-Waves" events far far away. The results of these calculations were not encouraging. Clearly, chances for detecting a signal that weak looked grim. On the other hand, these were "Guestimates" of the signal's strength. They were not certain by any means. If their estimates were wrong and there were objects producing much larger signals, then we would have missed a great opportunity. These kinds of arguments must have carried some strength with the funding organizations in Washington D.C., And the scientists making these arguments were hungry for funding to get their teeth into a real challenge. In other words, "Tell me it cannot be done, and we will do it!" Well, Weber convinced the funding organizations to fund him, and "once more into the breach," they went!

In this section, we will give an abridged account of some of the events.

First Attempt-Mod1 G-Wave Detector

Joseph Weber and his colleagues began their work on G-Wave detector designs shortly after returning from the GRF conference in 1957. The initial design, shown in Figure 4-10 A & B, consisted of a large aluminum detector ring cylinder – the antenna. The antenna was approximately three feet long and two feet in diameter, weighing approximately 1.5 tons as shown in Figure 4 -11A. When a G-Wave impinges on the antenna, a VERY SMALL part of the antenna's energy is released, in the form of a very slight motion of the atoms in the aluminum antenna. The signal detected by the antenna is converted into a very small electric current using a commercial motion detector.

This electric current is amplified by a low noise amplifier

(LNA). This process is repeated over other parts of the antenna to produce the largest possible signal. This configuration is referred to as “Direct Detection”. It is the simplest small signal configuration possible while being capable of detecting small signals. Even so, the direct detection approach proved to fall short by many orders of magnitude. The detector worked in principal, but the tiny detection signal was so small that it was swamped by the vibrations of other equipment. Even the ambient temperature of the equipment of the detector was enough to overwhelm the receiver with unwanted “noise’.

Figure 4-10 A &B: Joseph Weber’s first implementation of a direct-detection Gravitational Wave detector. The 1,5 ton aluminum nugget serves as the G-Wave detector element, moving as little as 1 Angstrom (10-10 meters) in response to a G-Wave illumination. Figure A is a photo of the device, while Figure B calls out the components.

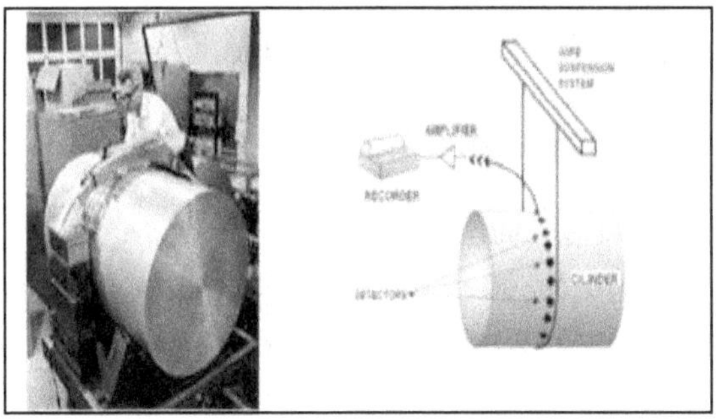

Weber’s team continued to work on their receiver to get the best sensitivity possible with their design. Around 1968, the detector was producing signals whose signature suggested that they could be called G-Wave candidates. The signals were coincidence types; that is, they were being simultaneously detected by both detector sights in the University of Maryland and Argonne National Laboratories, Illinois, at the same time. Weber announced his claim that their detection system was capturing the first G-Waves ever. The response from the scientific community; the claims were absorbed and analyzed for their accuracy and compatibility with the current knowledge of the structure of the universe and, more importantly, of the physics of the Milky Way Galaxy.

Following this initial response, other scientists began to design apparatus to reproduce Weber's published results. This is where the scientific community began to doubt his claims. Other scientists began questioning Weber's work and where the pile-on happened. The opinion of the community was that the work reported was probably not credible. Calls for more careful design and application of G-Wave detection systems began to emerge. So, again, we see the scientific method in action, paving a path toward an understanding of how nature works. The method is effective and reliable but not necessarily kind.

New Designs - The Interferometer

With the tenacity of a bull, Weber moved forward, along with many other scientists, seeking out means and methods like extracting tiny needles from a very large haystack. Over time, it became clear that a new method of detecting G-Waves was emerging; it is called Interferometry.

In Interferometry, a signal is split into two equal pieces. As shown in Figure 4-11, each piece is sent in a separate path from the other, reflected to its origin, and then the two signals are subtracted from each other.

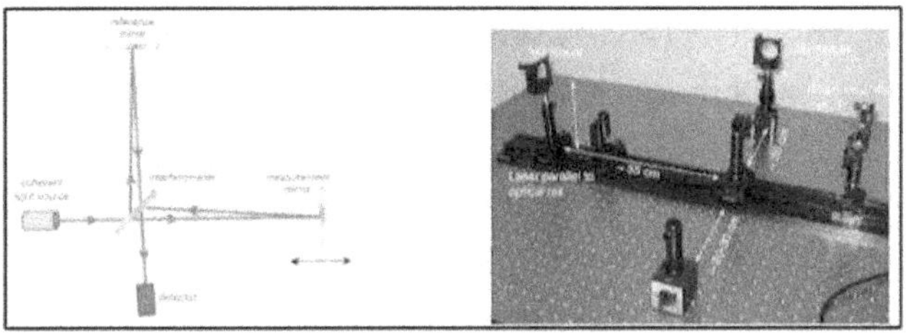

Figure 4-11: The basic layout of a Michelson Interferometer. Mirrors are used to split and steer the optical signal in the directions shown. The difference between the two path-lengths constitutes the detector's output signal.

The piece can be very small, and it is not subject to many of the noise sources of the direct-detect approach.

Michelson and Morley used the interferometer technique to measure the speed of light. The G-wave detector application of interferometry is complex, but its simplicity emerges in its application. The resolution capability is proportional to the length of the "arms' of the interferometer. This factor is critical to the success of this approach to G-Wave detection. In fact, as the design of the detector evolved, the size of the arms grew, some might say out of control, to the point where the arms had grown to four Km (2.5 miles) long! Needless to say, the LIGO (Laser Interferometer Gravitational-Wave Observatory) detector represents a serious pass at determining whether G-Waves exist. The concept was adapted by several researchers. It became clear that the interferometer was the tool that would make detection of G-Waves possible. The received signals were large enough to overcome the noise environment that prevented direct detection.

Meanwhile, the LIGO facility was becoming a highly respected research center for new detector concepts. By 1995, the funding organization, National Science Foundation, had invested $300,000,000 to support their research efforts. Finally, on September 14, 2015, the first Gravitational Wave was detected.

The three scientists, Barry Barish, Kip Thorne, and Rainer Weiss, were awarded the 2017 Nobel Prize in Physics for the discovery of Gravitational Waves at the LIGO research facility.

QUESTIONS FOR PONDER FODDER FROM CHAPTER 4

1. Name the five notable discoveries which contributed to understanding the nature of Big Space.

2. In your own words, describe the two theories of Relativity i.e., Special Theory and General Theory.

3. What is the Doppler Effect? How is it related to helping us know that the universe is expanding?

4. What was George Lemaitre's theory on how our universe began?

5. Do you think the Big Bang Theory explains adequately the origin of our universe?

6. How did the work of Penzias and Wilson support the Big Bang Theory?

7. What is a gravitational wave and if it exists, how would you measure it?

8. Who received the Nobel Prize for Physics in 2017 for the discovery of Gravitational Waves?

9. In your own words, why do you think scientific discovery costs so much in time and money?

Chapter 5- Small Space II

Quantum Behavior

Quantum Beam Splitters

Quantum Diffraction

Dual Nature of Light

Heisenberg's Uncertainty Principle

The Casimir Effect

The Nature of the Quantum

When a source of light shines brightly, as in our everyday lives, the light we see is wave-like. That is, light's behavior is almost entirely like other waves we experience, such as water waves or sound waves. For example, light from our sun fills the big space, which is our world. Water droplets in a rainstorm bend the sunlight to form a rainbow. Light passing through a window can be partially reflected off the window while the rest passes through the window. These are common everyday experiences that we all take for granted as we live our lives.

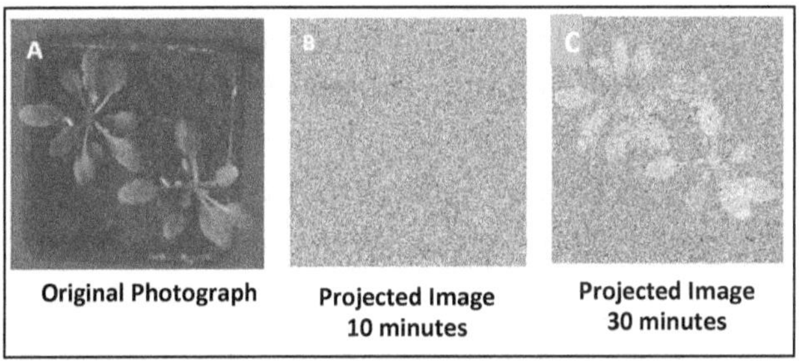

Figure 5.1: The corpuscular nature of light. As the original photograph is projected on a screen with extremely low intensity, we observe the graininess, produced by individual photons hitting the screen, building up the image as the exposure time unfolds.

But, when scientists shine the light that is exceedingly weak onto a screen from a source, we see something very different. We observe a random grainy pattern on the screen. The "grains" are all about the same brightness, but the weaker the light source is, the fewer grains we see. As we crank up the light brightness, the grain density increases. If we project an image onto a screen using very weak light and slowly increase the intensity, we see the image emerge from the random pattern, as shown in Figure 5-1. What we see is light impacting the screen, one grain of light at a time. These grains are called photons. Each photon carries the same specific amount of energy; as the intensity of the light beam is decreased further, the number of photons impacting the screen diminishes while the "energy per photon" does not.

This concept of a photon is one of the most fundamental in understanding what we know about the quantum nature of light. Figure 5-1 shows the formation of an image when a very weak light source is projected through the projector onto a screen (image A). After a few minutes of projection (image B), the screen has a grainy look to it because the photons are illuminating the screen randomly. After more time (image C), the image begins to appear. The random arrival of photons, over time, produces the same pattern we see immediately when the flashlight is emitting copiously.

Quantum Beam Splitter

Now, let's talk about another "experiment" we could do. We make a "partially reflecting mirror" by putting a very thin coating of silver on a glass slide. Light impingent on this glass slide gets "split" into two separate beams, a partially reflected beam and a partially transmitted beam (see Figure 5-2). We can make the mirror reflect 50% by adjusting the thickness of the silver coating. Then half the light is reflected while the other half passes through; that is, it is "transmitted." When there is plenty of light coming out of the flashlight, the two screens show a round spot, one for the reflected beam and one for the transmitted beam, as shown in the figure.

Figure 5-2: The beam-splitter has a partially reflecting surface so that half of the light wave's power is reflected and the other half passes straight through.

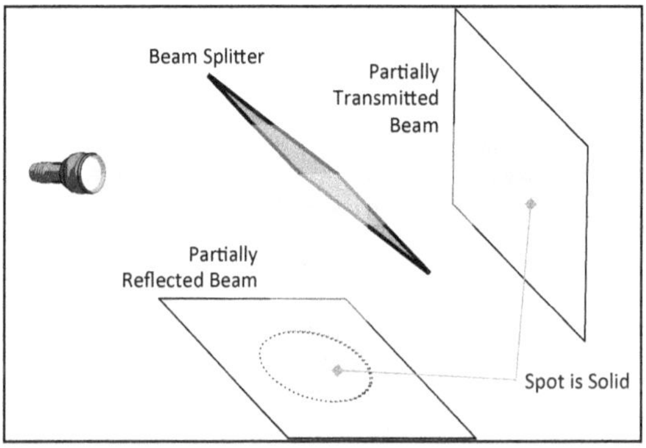

Next, in Figure 5-3, we attenuate the light from the flashlight until it is extremely faint but not totally extinguished. What we see then is that grainy pattern on both screens. If, as before, we replace the screen with a photographic film, we see the same pattern build up on each film until the round spots appear on each screen.

Figure 5-3: Quantum-limited Light. The beam splitter, when illuminated by faint light, divides the light into two equally illuminated beams, one photon at a time.

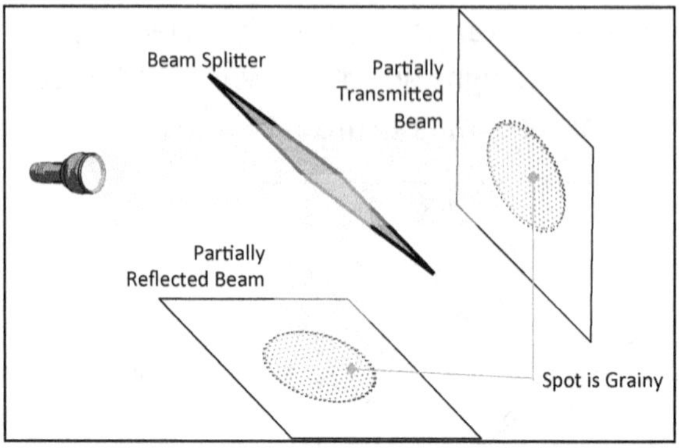

This may sound a little uninteresting, but let's back up a bit and consider what we just saw. Light is being emitted from the flashlight, one photon at a time and randomly. When each photon enters the beam splitter, it either passes straight through or gets reflected. It is not the same as the case of the brighter light from the flashlight. In that case, the light was a wave; half of the wave's energy passed through while the other half was reflected.

In this example, the light is not acting like a wave but a bunch of particles (wave packets) that are supposedly acting independently of each other. There is a lot of empty space between these photons, and the weaker the signal, the emptier space there is! Somehow, as these photons pass through the beam splitter, they seem to "know" which way to go so that, after enough of them have passed, half of them go straight through while the other half are "reflected." Are these photons talking to each other somehow, and if so, how? After all, they are each moving at the speed of light, so they would have to interact instantaneously, that is called "action at a distance", and that would violate the laws of special relativity that we discussed in Chapter 2.

With all our accumulated knowledge of the "rules" of quantum mechanics, we can use our equations to predict very accurately what will happen in the lab when we do these experiments, but truthfully, no one really knows why it works! How do we resolve this dilemma? How can light be a wave sometimes and a particle at other times, and even both at the same time in some cases? These issues have been discussed, worked on mathematically and have frustrated the brightest minds of science for a century and still they are not totally understood. As we will see in the next couple of examples, quantum behavior is more vexing than these simple examples would imply!

Quantum Diffraction – The Dual Nature of Light

So far, we have learned that light possesses many wave-like properties, such as the ability to transport energy through space as though empty space were its medium. We have also learned that light carries energy at a constant speed of 386,000 miles per second, independent of the location, speed, or acceleration of the observer. As with all waves, light waves vibrate at a characteristic frequency and have a characteristic wavelength. Green light has a longer wavelength than blue light and vibrates at a lower frequency than blue light. Light waves that vibrate at even lower frequencies are called radio waves or electromagnetic waves. If they vibrate at a very high frequency, we might call them "X-rays," but they still have a higher frequency. We might call them "gamma rays." All these types of waves are light waves of different "colors."

But we have also learned that under certain circumstances, light behaves like a particle. Historically, this feature was called the "corpuscular nature of light." When light behaves like a particle, its wave nature is somehow suppressed in lieu of particle-like properties. The photoelectric effect, as discussed in Chapter 3, was the first known example of this feature. Somehow, light simultaneously behaves as waves and as particles. Light has what is called in the physics community a "dual nature." This behavior is very difficult to really understand, and even though we can describe these features of light mathematically and quite often very precisely, there seems to be a level of understanding of light, matter, and gravity that still cannot be explained. As we progress in this text, we will be revisiting these important issues. But for now, we continue to explore the nature of the quantum.

Superposition Principle

There is a feature of waves we call the "principle of superposition." It is an intuitive concept but very important for understanding waves. If any two waves, moving in any direction, cross paths, then in the region where they comingle, they will pass through each other (See Figure 5-4.)

Figure 5-4: By the principle of superposition, waves pass through each other without colliding or interacting.

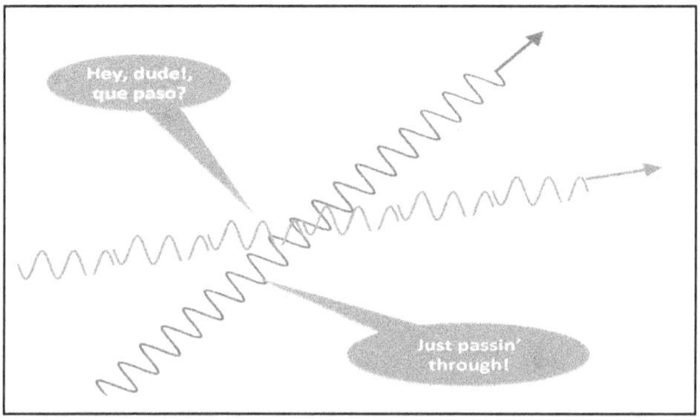

This means that in this region, as they pass through, their field values add together arithmetically. Once the two light waves no longer occupy the same space, their shape and features proceed unaltered. That is, waves superpose or pass through each other without interacting.

Even though waves can pass through each other, they can "interfere" with each other. This is a feature of the superposition principle. Waves "add together" as they pass through each other, rather than colliding or changing each other in any way. But when they occupy the same space, their additive properties can make them appear to bolster or diminish each other. Figure 5-4 illustrates this concept. When two waves' crests and troughs coincide, they are said to be "in phase." Under these conditions, their values add together, as shown in the figure, and the waves are said to interfere constructively (see Figure 5 -5A). If, on the other hand, the two waves' crests and troughs oppose each other (see Figure 5-5b), they interfere "destructively." It is not that the waves are destroyed; their values simply add up to zero in that case. Once the two waves no longer coincide, they move along as if they never shared the space. This is the principle of superposition.

Figure 5-5 (a) & (b): Superposition principle and interference. Waves that are "in phase" add together to make a bigger wave where they coincide. Waves that are "out of phase" cancel each other out. This property of superposition causes an interesting effect called "diffraction".

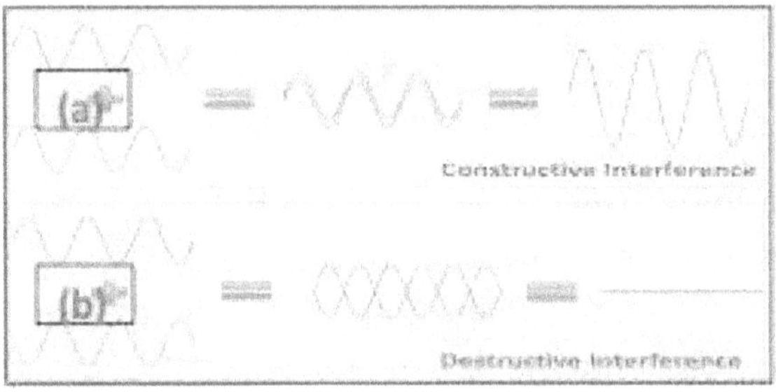

Wave Diffraction

Now let's talk about a well-known effect called "diffraction". What is "diffraction"? The average person has probably not heard of the word, but diffraction is a common occurrence in our everyday life. A working definition of diffraction is generally "the spreading of waves around an obstacle. The obstacle can be a doorway or the shutter on a camera. We often use the term "aperture" to describe the obstacle. This definition may sound very mysterious, but diffraction is a working example of the principle of superposition. When two or more waves of light are made to occupy the same space, as was shown in Figure 5-4 and Figure 5-5, they will "interfere" with each other to produce areas where the sum-total of light is large (constructive interference) and other areas where the light intensity is weak or null (destructive interference). This wave-spreading is often accompanied by a "diffraction pattern" that is directly connected with the shape and size of the aperture. A slit is one simple shape that was often used in early research to gain an understanding of the physics of diffraction.

Light sources like flashlights, indicator lights or large searchlights are all defined by both the size and shape of their aperture. In general, the smaller the aperture as compared with the wavelength of the light, the more spreading is observed emerging from the aperture. A simple example might be radio waves.

When driving our car through a hilly or mountainous region, we may notice that radio reception of FM signals may be obstructed by those hills or mountains. If the driver switches to an AM station, reception improves, although the choice of music may not! In general, this is because AM signals operate over a frequency range of 100 KHz which is much lower frequency than the FM band (ca. 100 MHz). The FM signal wavelength is much shorter than the hills, mountains, etc. (i.e., the obstacles or the apertures), so the FM signal bounces off the obstacles like billiard balls, obstructing reception. AM waves, on the other hand, are much larger (wavelength of two or three kilometers), so they will bend around; that is, they will diffract the mountains and reach the receiver.

> ***Apertures***
>
> *In future discussions on diffraction, the "slits" that let the light pass are more generally referred to as "apertures". An aperture's shape does not need to be a slit. While the use of a slit is most simple, the shape of the aperture can be more elaborate. Diffraction still occurs in all these cases, but the diffraction pattern will be different. In fact, there is a mathematical relationship, called Fourier Transform, between the shape of an aperture and the diffraction pattern it produces. This relationship is often exploited to produce "diffractive optical elements" (DOEs) that are useful in many applied disciplines such as medical imaging, signal processing and sensor systems.*

To understand this better, let's look in the lab at the progression of a light beam as it passes through a series of thin "slits." We begin with a beam of light moving strictly horizontally from left to right. We call such a beam of light a "plane wave" because the crests and troughs of the waveform are straight lines, as shown in Figure 5-6 (a). This beam confronts an opaque screen fitted with a single thin slit. If the slit, or aperture, is thin enough, about as thin as the wavelength of the light, the light emerges from the slit with the crests and troughs forming a circular wave pattern as shown. It is as though the slit was so thin that the emerging light appeared as a point source radiating its energy radially from the slit. This is the first instance of diffraction in our experiment. Continuing to follow the light from left to right in Figure 5-6 (b), the circular beam of light confronts a pair of slits. This "double slit" constitutes the equivalent of two point sources of light that can interfere with each other through the superposition principle; this is the second instance of diffraction. The resulting "diffraction pattern" appears on the screen on the far right of the figure. This understanding of diffraction was acquired in 1801 when Thomas Young performed the double-slit experiments for the first time. These experiments clearly demonstrate the wave nature of light.

Figure 5.6 (a) and (b) Single slit and double slit diffraction.

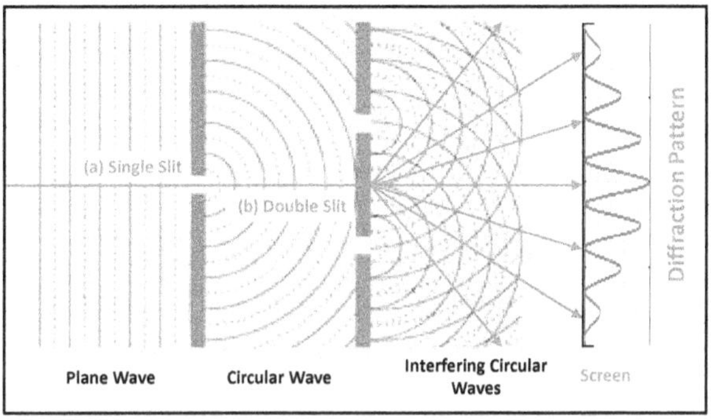

Diffraction is often described as the "selective bending of light", but it is more selective interference of light. The diffraction pattern on the right depends mathematically on the size and shape of the aperture.

So, with a basic understanding of what diffraction is and how it behaves in the classical world of large space, let's explore how particles act when passed through the diffraction apparatus. We will now consider diffraction in the quantum world. A word of caution: We are about to see our story of the quantum nature of light getting much stranger.

Light's Dual Nature

Let's now consider a very interesting, famous example of the dual nature of light. The dual nature of light can be and has been demonstrated in the lab many times. The process of discovery continues even today, with more and more examples found of the sometimes-strange behavior of small space. There is one set of experiments and observations, called the "double-slit measurements," that are most effective at highlighting the dual nature of light.

This set of experiments involves the diffraction of light and particles as well. It reveals the wave-nature of mass and the particle nature of waves.

Wave-Like Behavior

Diffraction of waves (water waves, radio waves, light, etc.) has been well-known and understood for more than a century. These measurements, first made by Thomas Young in 1801 established the wave nature of light. Referring again to Figure 5-6, we will first examine the diffraction of water waves or light waves to see what and how the waveforms, commonly called interference waveforms, are produced at the screen on the righthand side of the figure.

We observe the following:

1. The waves could be water waves or light waves. Water waves are produced by a ripple tank and light waves could be a laser beam or any other monochromatic, i.e., single wavelength.

2. The waves propagate as a plane-wave from left to right on the figure. They impinge on the single slit first, where they pass through the slit and propagate to the right as a circular wavefront toward the double slit arrangement.

There, two waves are produced, one by each slit. It is here where the two circular waves interfere. As shown on the figure, the crests and troughs interfere with each other to produce the wave pattern shown in the background of the diffraction apparatus.

For example, at the center of the light wave arrangement, two waves straddling the center line, and each other, propagate to the screen and recombine, constructively interfering to produce a large wave. At other points along the pool of waves, any two waves, that can travel different distances to the screen, will recombine at the screen, producing a larger or smaller wave depending on the difference in distances of the crests and troughs of the two waves.

At the screen is a plot of the wave height versus its position at the screen. This "waveform" is the signature of wave-like propagation.

Particle-Like Behavior

Now, we will look at what happens if we shoot macroscopic solid particles through the same double-slit arrangement. By macroscopic, we mean too large to fit in small space. We can imagine these particles as tiny spheres or even grains of sand. We also assume that these particles are strong enough to be unbreakable (see Figure 5-7. That way, we are always detecting an entire grain of sand. We first illuminate the setup with little particles, shoot them through the single slit, and then through the double slits Figure 5-7. To begin, we assume the particles are being emitted at a randomly high rate so that, at any instant, at least one particle is passing through a slit, and probably more than one.

We observe the following:

The particles enter and pass through the slits in an imperfect way; that is, sometimes a particle passes straight through the slits, but at other times it might hit the sides of the slits as it passes through the slit, causing the particle to be deflected to a new angle of departure from the slits. The particles continue to the screen, where they are detected and counted.

If we keep a tally of the screen arrivals versus position on the screen, we observe a pattern on the screen. It is what we might expect with solid particles shooting through large apertures. The pattern shows each aperture acting independently from the other, accounting for the marbles that sometimes ricochet off the sidewalls of the apertures. But there is no diffraction pattern from the macroscopic particles. Instead, the pattern is double lobe, with each lobe produced be each of the two slits.

We see no diffraction pattern, confirming the notion that a solid particle would not exhibit any wave-like behavior.

Figure 5-7: Double slit experiment done with small particles of sand, each grain indestructible and each grain as close to identical as possible. With particles, we see no evidence of wave-like diffraction. The arrival positions of these particles appear as a two-lobe structure with the lobes lined up with the two slits.

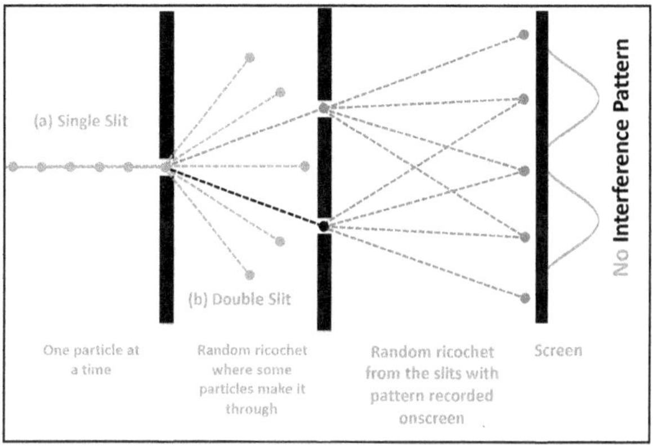

Electron Diffraction

Now we take a step into small-space. This time, instead of grains of sand or tiny marbles, we use electrons; these are particles with Small-Space dimensions; their mass is concentrated in a vanishingly small space, about 4×10^{-12} meters, or four trillionths of a meter - small enough? We allow the electrons to enter and pass through the slits randomly. As with the sand particles, we measure the screen arrivals of the electrons.

We observe the following,

Unexpectedly, we find that the pattern does not look at all like the pattern of two independent slits. Instead, we get a diffraction pattern, as shown in Figure 3-21, like that characteristic of the wave-like light wave or water wave patterns. This pattern is wave-like!

Could it be that two electrons passing through the slits are somehow interacting, or "communicating" with each other in the diffraction apparatus? If so, two electrons must be in the two slits for that to happen.

We ask what happens when we isolate the electrons that are being shot into the diffraction apparatus from each other. We can do this in the lab by arranging it so that the electrons are shot from the electron gun one at a time, with enough time between electron emissions to let everything settle down before shooting out the next electron. In this approach, only one electron is in the diffraction slits, evidently in one of the two slits at any time. So, we arrange the electron gun to emit single shots, one at a time, with sufficient time spacing that each electron arrives by itself, as shown in Figure 5-8.

Figure 5-8: Diffraction of electrons. The electrons as particles work their way through the diffraction apparatus, revealing their wave nature.

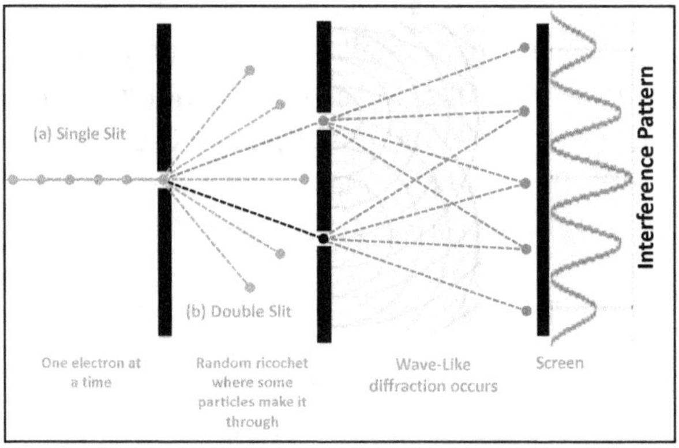

We observe the following,

After each electron is shot into the diffraction apparatus, a point on the screen lights up. As the number of electrons impinging on the apparatus increases, the screen begins to reveal a pattern, a diffraction pattern!

Somehow, even though only one electron is present, a

diffraction pattern appears.

This means that unlike marbles or billiard balls, electrons have a wave nature as well as a particle nature; that is, they have a "dual nature". The details of this dual nature remain a mystery, while scientists continue to experiment and to uncover more data to analyze.

This result is very strange. It is so strange that, even after nearly a century of pondering this set of experiments and creating more clever experimental techniques, there is still no strong explanation of what the heck nature is telling us!

But wait; there's more!

Heisenberg's Uncertainty Principle

So, particle-like marbles and grains of sand don't exhibit interference or diffraction in the lab, but electrons do, and so do photons. This last experiment, just described, left us with the disturbing realization that, somehow, one electron can be diffracted in the absence of other electrons. This is in direct conflict with what we would call wave-like behavior, and yet the result is wave-like behavior! What happens to generate this kind of result in the lab? Could it be that a single electron somehow splits up, with one half passing through the top slit and the other half passing through the bottom slit? Or could the single electron engage in meandering movement and manage to pass through the first slit first and then the second slit? These concepts are possible, but how do we measure this possibility? Without a measurement, we are without a data-driven answer.

Suppose that we modify the diffraction apparatus so that we can detect the passage of an electron through a slit? For example, let's imagine that each slit is outfitted with a very small light right at the departure side of the slit so that it is always shining light on the slit (see Figure 5-9). When an electron passes through and out of the slit, some of the light reflects off the electron at the instant the electron passes through the slit. This signal lets us know that an event has occurred and reveals in which slit it occurred.

So, we shoot an electron through the single slit into the double slit arrangement. In this case, let's say that the light at the top slit flashes, leaving a mark on the screen that indicates where the electron ended up. We continue this process, letting many electrons through the diffraction apparatus and tallying the screen results.

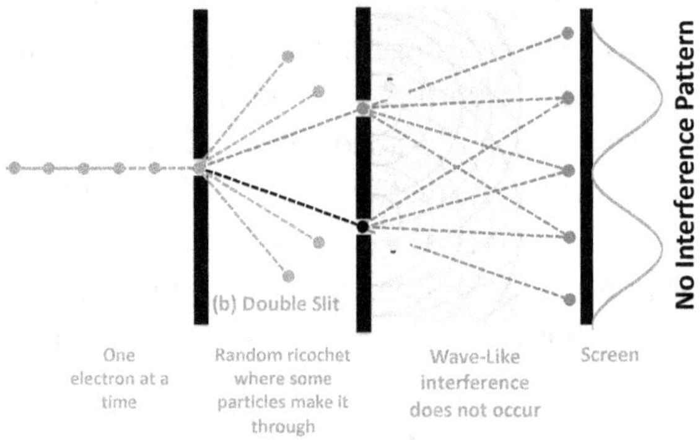

Figure 5.9: Heisenberg Uncertainty Principle.

We observe the following

After many electrons are shot through the diffraction apparatus, the fraction of events where the top slit is triggered approaches 50% and the number of events where the bottom slit is triggered also approaches 50%.

When we look at the screen, we see a two-lobed pattern we immediately recognize as indicative of particle-like behavior! The electrons that displayed interference and left a diffraction pattern on the screen, no longer show signs of interference; their wave-like behavior has disappeared!

Evidently, the act of shining a light on the slit as the electron is passing through the slit has caused the interference pattern to disappear. This phenomenon is an example of a more general effect called the "Heisenberg Uncertainty Principal". Applying the Uncertainty Principal to the diffraction of electrons, we can say "It is impossible to poll an electron that is passing through a slit in the diffraction apparatus without disturbing the electron and destroying the diffraction pattern." The more general Uncertainty Principal is a mathematical statement that says that the values of any two "conjugate variables" cannot be known simultaneously. This more complete and abstract description of the Uncertainty Principal is beyond the scope of this book.

The Uncertainty Principle carries great scientific and even philosophical weight.

In Classical Physics, i.e., prior to the advent of quantum mechanics, the state of any system was defined as knowledge of the position and momentum of every particle in that system. Such knowledge meant that at any other point on the timeline, one could ascertain the state of the system at any other time. In other words, determinacy was assured if the state of the system was known. Philosophically, this meant that we could all rest assured that our world was predetermined. The development of quantum mechanics and the Uncertainty Principle evaporated the ideas of such determinacy. To some people, this situation is very troubling. However, it now appears that we live in a world that is not as certain and clean-cut as some would have it.

The strange and unexpected results of the double-slit experiments ignited a series of spirited discussions and debates among the heavy hitters of physics, including Einstein, Niels Bohr, Paul Dirac, and Paul Ahrenfest, just to name a few. In particular, the interchange between Einstein and Bohr was very intense. These two physicists disagreed on the meaning of the results of the diffraction experiments. Einstein's view was that the quantum mechanics as it was in ca. 1933 must be incomplete. His view was that quantum mechanics that embodied such uncertainty cannot be expected to fully define the state of a system. Therefore, the quantum theory of 1933 must be incomplete. Niels Bohr, the father of the quantum mechanical model of the atom, held the opinion that the uncertainty in a quantum state was, to date, being confirmed in the lab.

Bohr was an experimentalist, and Einstein was a theoretician.

Little did they know that the debate would continue unabated to our present time, and who knows how much longer!

> ***Physics and Mathematical Intuition***
>
> *It would be easy to believe that intuition and scientific method are incompatible with each other. After all, the rigors of experimental verification, insistence on reproducibility of results and peer review don't seem related to intuitive flash-thinking. However, it is not at all uncommon for a scientist, after years of long hours of experimental or theoretical research to develop an intuitive sense of how nature works. Occasionally the scientist might develop a "hunch" or "gut feel" on how a certain physical process might work. When this happens, the scientist will construct a theoretical model of the process and describe the expected outcome of experiments. From that point, if the theory appears plausible, scientific method is used to play out the process and determine the validity of the intuitive idea. The intuitive hunch may have initiated the idea, but the Scientific Method provided a rational framework from which the validity of the hunch could be tested.*

Casimir Effect

We have been talking about small space, describing how small space can support wave propagation as though Small Space were a medium. We talked about how particles and photons can be created at will (whose will?) or destroyed at will. Many of the newly discovered "particles" we hear about in the news, like the Higgs' boson, are better described as "resonances": they can be born out of empty space, exist for a very short time as defined by the Uncertainty Principal, and then they spontaneously decay by returning to "empty space" as pure energy. With lab measurements revealing this kind of particle physics, it may not be unreasonable to imagine space as not so empty after all.

Figure 5-10: The large yellow atoms are held in suspension by the Brownian motion of the smaller blue atoms in the colloid.

If Small-Space is in fact not so empty, then can the contents of empty space be seen or somehow detected? Well, this sounds like a tough challenge, but in the physics community, there is always someone looking for a problem that seems impossible to solve. If there really is content in Small Space, then it might be possible, difficult, impossible-–no, to detect it. Well, it happens that there was someone who welcomed the challenge and who had a game plan to succeed. His name was Hendrik Casimir. It is an interesting story, and it is quite fantastic but true. The story begins around 1940 or 1942 at Philips Research Labs at Eindhoven in The Netherlands. A very promising research scientist named Hendrik Casimir had recently been hired by Philips Labs to do some applied research on colloids. Colloids are gels, emulsions and other goopy stuff that is neither solid nor liquid, but something in-between. They are useful in the manufacture of paints, inks and thickening agents that are used to control the viscosity of liquid materials like paint, for example. These materials are typically made from very large molecules that are suspended in a second often liquid material, which produces the aggregate colloid material.

Hendrik Casimir

Discoverer of the Casimir Effect

Born July 15, 1909

Died May 4, 2000

Danish

Wilhelm Exner Medal (1982

Matteucci Medal (1985)

One of the Founders of CERN

To design colloids with predictable behavior, however, the inter-molecular forces, called van der Waals forces, that keep these large molecules working together must be well understood, but they were not. In fact, the existing London-van der Waals theory did not agree at all with experimental results. Casimir believed intuitively that there was an error in the way the theoretical physics of large molecules was being modeled. He believed that the problem could be resolved by employing a newly developed mathematical theory called "Quantum Electrodynamics," or QED. QED is a mathematical theory that describes how electricity and magnetism work in the quantum world;-- that is, in Space. While the scientific and mathematical details of how Casimir resolved this issue are far beyond the scope of this text, this is still an interesting story.

In 1940, Hendrik Casimir was a young 31-year-old theoretical physicist. A highly accomplished scientist, he had already been employed as a faculty member at the

University of Leiden in The Netherlands. He was also the Managing Director of the Kamerlingh Onnes Experimental Physics Laboratory of the University of Leiden, where his wife, an experimental physicist, also worked.

Despite this impressive career trajectory, Casimir's life was in turmoil and about to change. In March of 1940, Nazi Germany invaded and occupied The Netherlands, closing the University of Leiden's Onnes Lab and leaving Casimir both married and jobless. Because Philips Lab was doing research on large molecules and the colloid problem, they offered Casimir a research position at their lab. Casimir had found an engineering problem to work on, one that called for the state-of-the-art in theoretical physics to get the job done. This was important research! He accepted their offer.

The Physics

While researching the physics of colloids, Casimir realized that the current theory of the forces holding large molecules together was flawed, and that flaw could produce large errors in calculating the Van der Waals force holding these molecules together. Casimir and a couple of colleagues, including Niels Bohr, speculated that the Van der Waals calculations failed to account for the speed of light; that is, they assumed that the speed of light was infinite. From a theoretical physics perspective, neglecting the speed of light in the calculation was a reasonable assumption since the distances between atoms were very small. However, including the speed of light was most properly accomplished by using a mathematical formalism called "Quantum Electro-Dynamics" (QED) instead of classical electrodynamics to perform the calculation. It was through this change of mathematical process that

Casimir realized that he could use the same kind of calculation to determine the forces between atoms and molecules at the microscopic (or Small-Space) level. This all seemed reasonable to him, so he proceeded to calculate. One of the consequences of using QED to calculate Van der Waals forces was that doing so would also bring along the mathematical content of the vacuum state, or "zero-point energy," into the calculation. Even at absolute- zero, it manifested the energy content of the vacuum or empty space.

Casimir's calculations were successful in modifying the Van der Waals force to agree with measured results. They resolved a long-standing problem with the theory of the Van der Waals force. This success strengthened Casimir's understanding that "empty space" is somehow not empty. Casimir understood that the necessity of including QED to successfully compute Van der Waals forces meant that there must be a strong interaction between these forces and the vacuum (i.e., Small-Space). This motivated him to calculate the force exerted on a test-mass in an otherwise empty volume. He performed these calculations and reported the results in an often-cited paper.

More on Atoms as Resonators....

Because the resonator is compatible with the simplest description of an atom, it has often been used as a basis for modeling energy bearing media.

Like an atom, the resonator can receive energy, store energy, and emit energy. When used with a quantum mechanical description of the resonator, the system offers a simple but reasonable estimation of an atom.

For example, when an atom absorbs energy from space, for example a light wave, it stores and re-emits energy according to certain rules of quantum mechanics. Conversely, when an atom emits energy into space, it changes to a lower energy state that is compatible with the emitted energy.

Detecting the Vacuum

The results of Casimir's calculation implied that the effect of the "vacuum fluctuations" would be very much present, but their signature signals would be very small, right on the edge of detectability. This assessment may sound discouraging, but for a physicist, it constitutes more of a challenge than discouragement. As we have noted before, the only way to confirm a theory is through measurement and demonstration of reproducibility of the results.

To get a better feel for the experimental challenge of such a measurement, let's consider an enclosed volume of "empty" space. As shown in Figure 5-11, that volume is contained in a thin, a few microns layer between two parallel plates. Applying our knowledge of standing waves to an imaginary volume of space, we arbitrarily assume that our empty space is "filled" with resonators. Remember, we discussed atoms as resonators in Chapter 3.

To simulate the presence of resonators in empty space, we can imagine an array of such resonators producing standing waves. Each resonator has its own set of vibration modes, depending on the resonators' geometric configurations. Figure 5-11 shows a functional layout of a set of parallel plates, separated by a small distance, typically one micron or less. The space between the two plates is populated by a set of resonator modes. This set of modes is different than the ones in the surrounding space. Even though the set of plates is in thermal equilibrium with the surrounding space, some of the vibrational modes in the parallel plates cannot be supported because the wavelength of these standing waves is too long (i.e., longer than the length of the parallel plates). Therefore, the two plates will tend to move towards each other to fill the space that had been occupied by the empty space. In less technical language, the plates are attracted because "Nature abhors a vacuum."

However, the theoretically expected results of such a measurement were quite small. The inward force expected was approximately 2 x 10-4 Dynes, or 5 x 10-10 Pounds.

Figure 5-11: If two parallel plates are placed very close to each other, that enclosed space will support fewer modes of vibration than the free space. The plates will experience a greater inward force than outer force, resulting in a net "attraction" between the two plates. https://byjus.com/physics/casimir-effect/

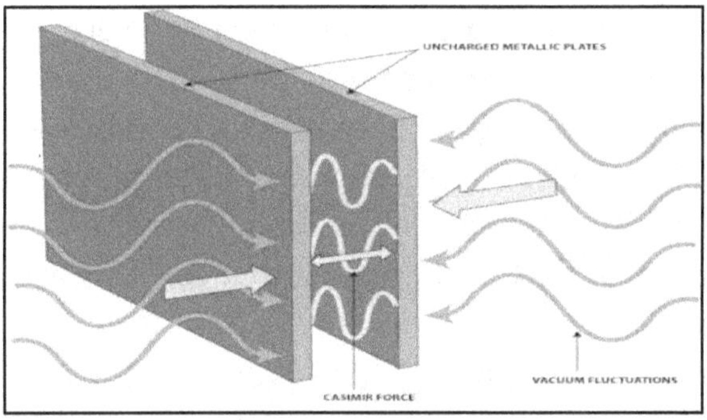

This is nearly a vanishingly small result, but a clean measurement would likely change the perspective of the importance of empty space and the roles that are played by "zero-point energy" in quantum mechanics.

The first reported result of measurement of the Casimir effect was by Marcus Sparnaay et al. (Philips-Eindhoven, 1958). This measurement revealed its complexity and difficulty. The first results were not at all conclusive, but within the measurement errors, they were not inconsistent with the expected results. Over the next several decades, numerous attempts were made to measure the Casimir effect. Each time, clever measurement techniques were used, resulting in cleaner and more reproducible results. One of the landmark measurements was reported in 1997 by Steven Lamoreaux et al. at Los Alamos National Laboratory. The measured results were within 5% of the theoretical predictions. Since that time, the results have been reproduced many times at different labs.

Evidently Empty Space is not so empty after all!

What do we mean by this?

In discovering and ratifying the Casimir effect, we have established that Empty Space is not so empty. The reader may already realize that empty space is not so empty, but the implications of this are far-reaching. In a sense, we have gone full circle. We began by assuming that there must be a medium as part of space, one that carries waves of electromagnetic or gravitational energy from one place to another. We saw how measuring the speed of light revealed that there is no inertial reference point in nature, and that implies that there can be no "ether," which in turn means that empty space must be empty.

This "cosmic dilemma" begs for a solution. As much as scientific endeavor advances, these issues of reference frame consistency persist. Isn't nature a beautiful mystery?

Questions and Ponder Fodder for Chapter 5

1. How do we know that photons exist?

2. What is the Superposition Principle?

3. What is diffraction? Name an example of superposition from classical physics.

4. Why is it that a 750 KHz AM radio signal will generally propagate much further than a 2 MHz FM radio signal?

5. What scientist first demonstrated the dual nature of light?

6. What is meant by the "dual nature of light?

7. Name a behavior of electrons (i.e., particle) behavior that falls into the dual nature of particles.

ABOUT THE AUTHOR

Davis Hartman was born in 1947 in Pennsylvania and moved to New Jersey at 10 years old. His undergraduate education was at a liberal arts school named Upsala College in East Orange, N.J. While there, he received his first taste of real Physics. When he was 18, he was assigned to a research position at the Penn-Princeton particle accelerator at Forrestal Labs. He recalls, "It was the first time I really felt alive! I was convinced that physics was for me."

After graduating from Upsala, Davis enrolled at the University of Arizona in Tucson, AZ, and declared his matriculation goal to obtain a PhD in Physics. While there, he moved his research interests to cosmic ray Physics (a branch of particle physics). After seven years, he left U of A with his PhD in hand.

After graduating, he accepted an employment offer from Bell Laboratories, where he engaged in research and development (R/D) in optoelectronics and fiber optics. A few years later, he moved to Motorola, where he engaged in R/D in next-generation ICs (integrated circuits). The next stop was Bellcore research labs, then back to Motorola for a few years, and finally, seven years of R/D at General Dynamics. During these years, as an experimental physicist, he did research and learned how it was done (or not done....!). He observed the operation methods of moving research results from the lab to the marketplace.

Davis, in his career, has had numerous achievements:

He was part of the Ocean Systems Research Team at Bell Labs, which designed next-generation fiber optic telephone systems that revolutionized intercontinental communications.

While at Bellcore, he was a research fellow at Hitachi Corporation in Japan.

He holds 54 issued patents, has published widely, and is believed by some to be the "father" of short-reach computer interconnects, which enhanced computer systems' performance.

He has contributed to research on our defense systems, but most of his work has been in the commercial sector.

All the above are some of Davis's many contributions to physics. However, anyone with the privilege of knowing or working with Davis knows that his work does not define him. He would say, "Physics is what I do but not who I am."

He is most proud of being a good husband, father, grandfather, and friend.

Daivs currently resides in Arizona, and can be reached at:
davishartman548@gmail.com

Acknowledgements

Writing this book has been an extraordinary journey, one that would not have been possible without the encouragement, support, and wisdom of many individuals.

To my wife, Sara Crosby Hartman, your love, patience, and unwavering support have been my greatest source of strength. Thank you for always believing in me and for standing by my side through every challenge and triumph.

I would like to express my sincere appreciation to Dr. Michael Lebby, Paul Baker, Tom McMullen, Alex Crosby and his dad, Wayne, John Mattos, and Melody Kelly for taking the time to read this work and offer their thoughts. Their support and interest in this project have meant a great deal to me.

A special thank you to Julia Rogers and Lara Hartman for their keen editorial insights and for ensuring that the complexities of quantum physics and cosmology remain accessible and engaging to readers from all backgrounds. Your dedication and attention to detail have been truly invaluable.

To Deborah Rychel, I am deeply grateful for your meticulous work in copying and editing, ensuring that this book is as polished and precise as possible.

To Michelle and John, thank you for supplying the environment of the Breakfast Bar in Boise, the place where the original idea for this book was triggered.

To my family, friends, and colleagues—thank you for your patience, support, and for always inspiring me to explore the unknown. This book is a testament to the collective curiosity and relentless pursuit of knowledge that we all share.

Finally, I extend my heartfelt gratitude to those who have supported me over the years, including those whose names I may have unintentionally omitted.

www.ingramcontent.com/pod-product-compliance
Lightning Source LLC
LaVergne TN
LVHW022052200926
848481LV00003B/879
* 9 7 9 8 2 3 1 6 3 8 9 2 5 *